Insects of
the World

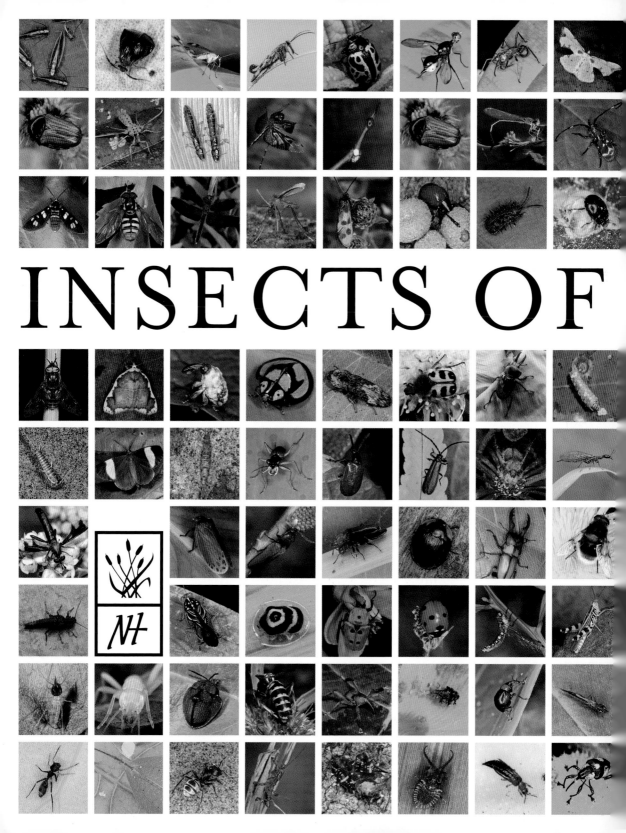

INSECTS OF

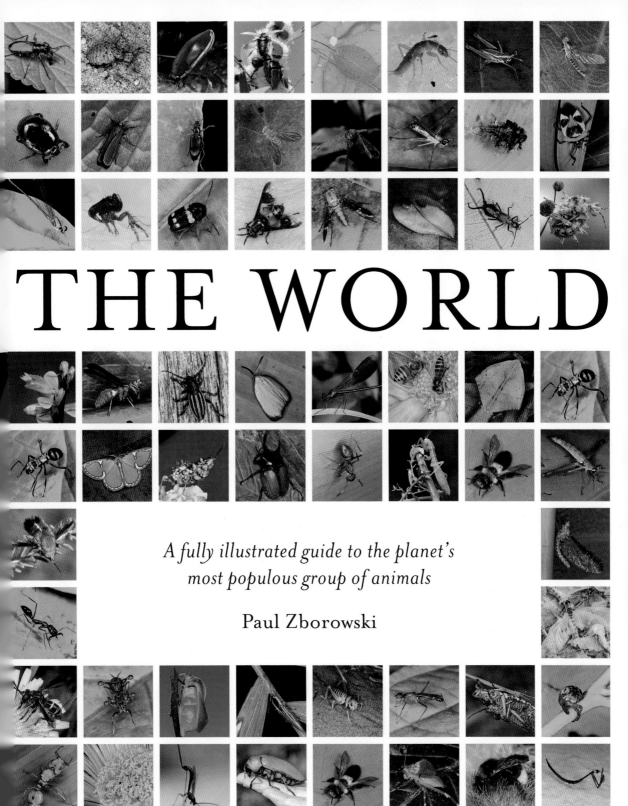

THE WORLD

*A fully illustrated guide to the planet's
most populous group of animals*

Paul Zborowski

A mass of butterflies descending onto Amazonian riverbank sand, to gain an essential bounty. Rare compounds such as urea, which male butterflies require, are often found on riverbanks.

CONTENTS

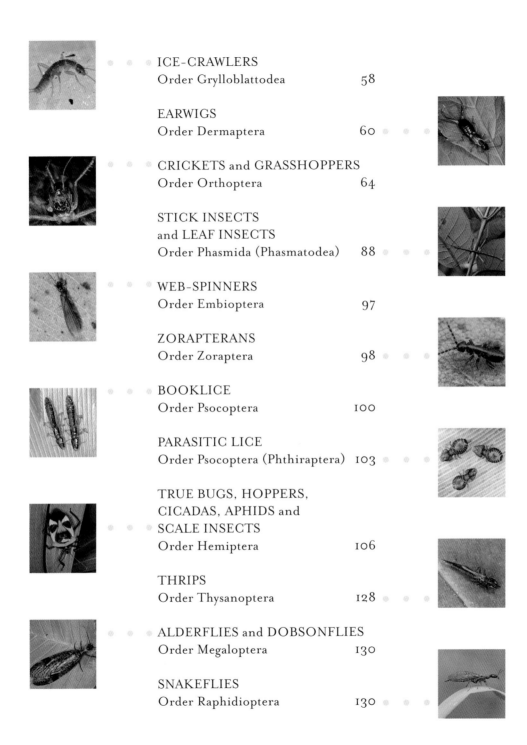

INTRODUCTION

Insects are quite literally the largest subject in the natural world. Where to begin and end a book that hopes to introduce, enthuse and teach about this endlessly fascinating subject?

With more than one million species known, and at least as many awaiting discovery and naming, the stories each species represents are close enough to endless. Using the simple observation that every picture is worth a thousand words, the 1,000 plus images in this book are all presented with a caption that starts the process of discovery. When so many different lifestyles are possible, simple facts about what a particular insect does, why, where and how, can turn into novel and often weird tales.

Presenting insects from all over the world allows a showcase of many different habitats. Insects have adapted to all the major habitats on Earth, and a few very extremely surprising ones. They are herbivores, hunters, fungivores, detritus-eaters, parasites of each other, and sometimes of us. While it is a common adage that the rainforests have the most species, a typical house has a surprising number too. Even the driest deserts have insects adapted to survive the near impossible. There are insects that despite being cold-blooded, live on top of snow-covered landscapes. Flies that dance and hunt on the surface of hot springs, and insects which never see the light of the sun, living deep in caves and underground rivers.

Evolution

Insects have been around since long before the dinosaurs, with the earliest known fossils dating back nearly 400 million years. The fast turnover of generations, and the vast numbers involved in each generation, provide the opportunity for random mutations to sometimes be beneficial for survival. A gene allowing better sensory information, or longer hunting reach or speed, is likely to prosper in further generations. But what is most amazing about the large possibility of expressed random mutations, is what this does to the appearance of insects. Some very showy, wildly bright, cryptic, or very spiny, or hairy, insects may be expressing a trait particularly useful to survive. But the general oddness of shapes and patterns may simply represent traits that are not disadvantageous. If one of these expressions of a gene does not actually disadvantage the species, it can stay in some, or all populations. Hence the answer to why insects are so diverse and surprising is not always a logical one, but simply a neutral one. It neither helps

A well-preserved fossil of a large dragonfly from about 140 million years ago. Note the larger-than-usual wings. These insects shared the planet with the dinosaurs, which at this time, the Upper Jurassic, were at their most dominant. However, earlier proto-dragonflies, monsters with 70cm (28in) wingspans, date back to twice this age.

nor hinders, so it persists for now.

And then there is the joy of deception. Reds, oranges and yellows are generally perceived by predators of insects to be warnings. Very bold contrasts of reds, blacks, yellows and whites for instance, often mean that an insect has been eating poisonous plants and retained the poison in its body. A bird eating such an insect usually spits it out and does not die, but it learns to associate the image with danger. However other insects, which are not poisonous, sometimes evolve similar patterns, and thus get some protection from this ploy, which is known as mimicry. The game is a fluid one, as too many mimics will not allow predators to learn the lesson of danger often enough, and they get eaten anyway. This constantly changing game of natural selection at least partly explains why there are so many forms and so many species. Fragmenting habitats, isolating populations from each other, environmental changes (such as global warming) and sudden catastrophes are more ways species change.

This book presents the insects in an evolutionary order, starting with the oldest ones such as the silverfish, and ending with the most recent line, the bees.

The Insect Body

Insects are part of a huge phylum, the Arthropoda, which roughly means 'jointed-legs'. These joints are in all body parts, covered by a hard exoskeleton. This skin replacement is so tough that the creatures need no internal skeleton.

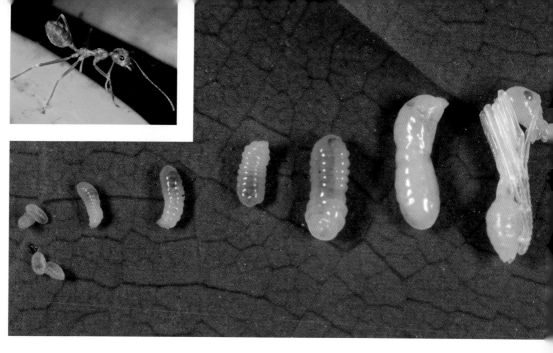

The complete metamorphosis lifecycle is one of the most remarkable adaptations of insects. This ant, *Oecophylla smaragdina*, shows every stage from egg to larval growth stages, and the pupa (right), where the body is rearranged into the adult stage. Such a system allows two different lifestyle strategies in one species.

Because they are very light, this protection is generally good enough to survive a fall from any height, and a lot of ritual and real fighting. The exoskeleton can also be soft, as in caterpillars, but even this is still made from the same superb protein called chitin. It is not just tough, but water resistant, so that, for instance, insects can survive in hot deserts without instantly drying up.

There is no lung, or air pump equivalent, in insects. Instead they possess breathing 'holes' called spiracles along the sides of the body. These diffuse air into the body and sometimes can be closed, especially in aquatic insects. This system works well, but is also one of the limitations to the overall size of insects. Size variation is extreme, from micro parasitic wasps which are all but invisible at less than 0.4mm (0.015in) long, to beetles 16cm (6.5in) long and weighing more than some small mammals. However, the girth of the body is limited by how much air can be diffused in. Thus the longest insect in the world, a skinny-bodied stick insect, can attain perhaps 60cm (24in), but the fat-bodied goliath beetles in Africa do not exceed about 6cm (2.5in) width. Some long-distance fliers, such as the locusts, have bodies soft enough that the action of the beating wings produces a pump-like flexing of their thorax, forcing more air in.

The body plan that defines insects is composed of three main areas: the head, thorax

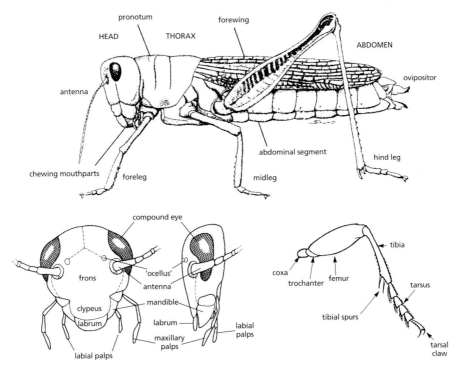

A typical adult insect – a grasshopper – illustrating the major parts of the body.

and abdomen. These sections in turn are made up of many separate segments and plates. When the insect needs to grow, it sheds the whole exoskeleton, at a time when the next one is ready to be exposed and harden on contact with air. The process is known as moulting (or molting). Descriptions of life cycles often talk about how many moults does a larva or nymph undergo during development. This normally happens at night, as the insect is vulnerable during this operation.

Insects which do not have a pupa stage, look like small adults when born, and shed their 'skin' from 5 to 20 or more times. Only on the last shed, or moult, do they develop wings and sexual parts, ending in the adult stage. Grasshoppers are a good example, as the tiny wingless hoppers eventually develop wings. This is called incomplete metamorphosis.

Insects which do have a pupa stage undergo what is called complete metamorphosis. They have a larva, or caterpillar stage, which is essentially a grub-like eating machine. After six or eight moults they spin a silk cocoon or shiny pupal case. Inside they completely rearrange their cells to become the usually winged and functionally different adult. Butterflies and beetles are the best examples of this. This type of development

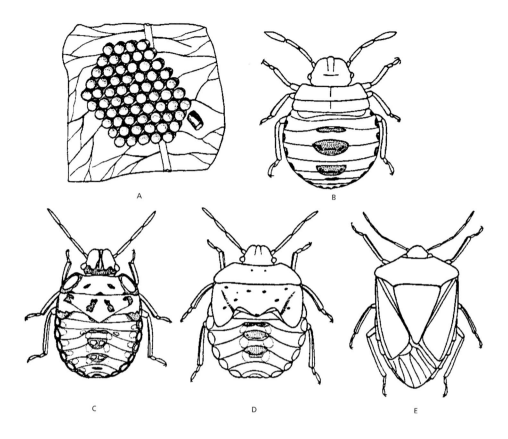

Stages in the incomplete or exopterygote life cycle illustrated by a bug of the family **Pentatomidae** (order Hemiptera): (A) eggs; (B) first instar nymph; (C) third instar nymph with very early wing buds; (D) final instar nymph with wing sheaths; (E) the adult or imago, with full wings.

has the advantage of two separate lifestyles, different food requirement, or even no food needed for a brief adult stage.

There are many hundreds of names for the many body parts of insects. Here we introduce a few, which apart from being a starter, will allow the image captions in the book to be able to tell more detailed stories...

Classification

The practice of describing new species, and placing them into the groups they are related to, is called taxonomy. Every new species found fits into a vast family tree. Currently insects, the class Insecta, are divided into 26 orders. These are the big divisions that

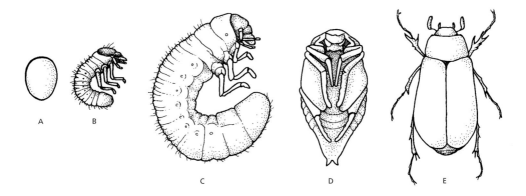

Stages in the complete or endopterygote life cycle illustrated by a beetle of the family **Scarabaeidae** (order Coleoptera): (A) egg; (B) early larval stage; (C) final larval stage; (D) stationary pupa showing developing adult characteristics such as wing sheaths; (E) winged adult or imago.

separate, for instance, the beetles from the flies. Within these orders are families. These separate, for instance, within the beetle order, ladybird beetles from scarab beetles. And within these families are genera and species. A scientific name for any creature uses both the genus and species name, often derived from Latin and Greek words that help describe it. For example, the Seven-spot Ladybird beetle is called *Coccinella septempunctata*, with the Latin species name translating simply as 'seven spotted'. Note also that species Latin names are published in italics.

If you know the family of an insect it tells you something of its life cycle and probable habits. This book, which is mainly a pictorial tour through the insect world, identifies all insects to at least their families. Most are named to genus and often to species. All the insects are photographed in their natural habitats, all over the world. This shows how they stand, eat, hunt, camouflage, and more, allowing for each caption to tell a story. However, as they were not taken as specimens for later analysis, full identification was not always possible. As many were found in very exotic corners of the globe, some may represent species not yet known to science, or awaiting names. Every chapter starts with numbers of families and species for that order. These numbers are in a state of flux. They continually evolve as more species are discovered, or as modern molecular scientists tinker with the more practical, existing morphological (visible), taxonomic divisions.

Please enjoy the myriad stories in the picture captions, and the amazing variety of these beasts. Hopefully it will serve to inspire further interest, therefore a comprehensive multi-format bibliography at the end will steer you deeper into this world.

Class Insecta

Sequence of Orders

PRIMITIVE, WINGLESS INSECTS – THE APTERYGOTA

Order Archaeognatha	bristletails
Order Thysanura	silverfish

WINGED INSECTS (ADULT STAGE) – THE PTERYGOTA

Exopterygota – incomplete metamorphosis

Insects with immature stages (nymphs) that are similar to the adults and develop their wings on the outside. Such as the cockroaches, which start out looking like tiny adults, and develop wings only on the final moult to the adult stage.

Order Ephemeroptera	mayflies
Order Odonata	dragonflies and damselflies
Order Plecoptera	stoneflies
Order Blattodea	cockroaches and termites
Order Mantodea	mantids
Order Grylloblattodea	ice-crawlers
Order Mantaphasmatodea	heel-walkers, gladiators and rock-crawlers
Order Dermaptera	earwigs
Order Orthoptera	crickets and grasshoppers
Order Phasmatodea	stick insects and leaf insects
Order Embioptera	web-spinners or embiids
Order Zoraptera	zorapterans
Order Psocoptera	psocids, booklice and parasitic lice
Order Hemiptera	including true bugs, hoppers, cicadas, aphids and scale insects
Order Thysanoptera	thrips

Endopterygota – insects with complete metamorphosis

Insects with immature stages (larvae), which differ from the adults, and that undergo a pupa stage where wings develop on the inside. Such as the caterpillar–pupa–adult sequence of butterflies.

Order Megaloptera	alderflies and dobsonflies
Order Raphidioptera	snakeflies
Order Neuroptera	including lacewings and antlions
Order Coleoptera	beetles
Order Strepsiptera	stylopids (small insect parasites)
Order Mecoptera	scorpionflies
Order Siphonaptera	fleas
Order Diptera	flies
Order Trichoptera	caddisflies
Order Lepidoptera	moths and butterflies
Order Hymenoptera	wasps, sawflies, ants and bees

BRISTLETAILS

Order Archaeognatha

470 species in 2 families

The bristletails are a very seldom seen group of primitive insects. They are superficially very similar to the silverfish (next chapter) but differ by their bodies being arched, not flattened, and the central of three 'tails', or cerci, at their rear, is longer than the outer ones. The head sometimes has very large eyes, meeting in the middle.

There are about 470 species worldwide, which hide in leaf litter, rock crevices, under bark, beach flotsam and in wet forests. Most are nocturnal and scamper very quickly if disturbed in the light. Some even jump away with rapid flicks of the body. Their food is mainly decaying vegetable particles, algae and lichens. After mating, small batches of about 15 eggs are hidden. These hatch into adult-like larva, which mature after about nine moults.

Right: The commonest place to spot elusive **bristletails** is on rocky beaches. Crevices above the high-tide mark often hide large numbers of these insects, feeding on algae and plant detritus brought in by the sea. This grey species is common in Europe. 1.2cm (0.5in).

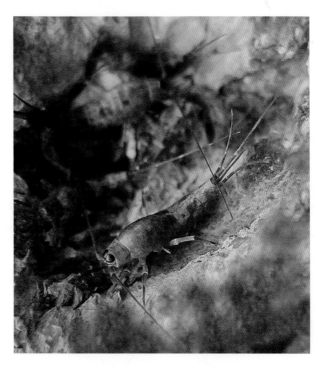

Opposite: A rainforest **bristletail** in Borneo. The gloomy moist habitat allows this species to be active in daytime, browsing on the algae on tree bark. 1.5cm (0.6in).

SILVERFISH

Order Thysanura
(Order Zygentoma)

400 species in 4 families

These ubiquitous ancient creatures have not changed their very well-adapted bodies for more than 300 million years. Their appearance is similar to that of the bristletails, although the three 'tails' are of even length, and the body is not arched but flattened. Most species also have small eyes, unlike the sometimes huge eyes of the bristletails. There are about 400 species in the world, which hide in leaf litter, under bark and in caves. Some of them, in the family **Lepismatidae**, have adapted perfectly to live in human dwellings. These are the classic silverfish, so called as their bodies are covered in silver scales. The minute scales are very slippery, and freely dislodge when rubbed. This is an excellent defence mechanism as most potential predators come away covered in scales, while the silverfish rapidly escapes. Even a normal spider web would only catch scales, except for one specialist predator, the spitting spider. These small free-living spiders have combined the venom gland with a silk gland, and fire or spit short lines of silk hundreds of times in one second. Even the slippery silverfish winds up pinned under a net of web.

These secretive creatures are long lived, up to four years, and tend to be active only in the dark. The household species can be pests as they eat glues and some clothing materials, including silk. In the wild other species are omnivorous, feeding on decaying animal and plant matter. Some species have adapted to live in ant nests, coexisting by imitating the smells familiar to the ants.

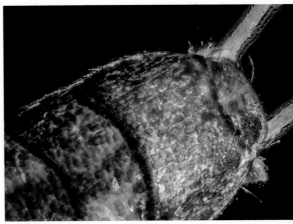

The classic **silverfish**, *Ctenolepisma* sp., which is common around the world in our houses. Here it is in its pest element, sitting on a book. The starchy book-binding glues are a favourite food. 1.2cm (0.5in).

A close-up view of the scales of a **silverfish**. Not only does it show where the name came from, but it is easy to imagine these slippery loose scales coming off when a predator tries to grab it.

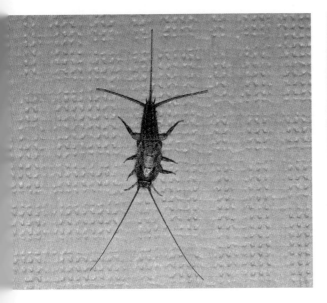

The other name these insects are known by is **firebrat**. *Thermobia domestica furnorum* is common in warm houses, where it is attracted to the warmest spots, such as the boiler rooms of centrally heated dwellings. It is less of a pest, but very hard to tell apart.

One species of **silverfish**, in the same genus as the household pest *Ctenolepisma*, lives in the rain-less dunes of the Namib desert in south-western Africa. A variety of specialist insects have evolved here, taking advantage of morning fogs rolling in from a cold Atlantic current. Before sunrise the silverfish digs a furrow on a dune side facing the fog, and drinks the water condensing on the edges. 1.5cm (0.6in).

MAYFLIES

Order Ephemeroptera

3,000 species in 42 families

Mayflies are familiar to anyone who goes fishing as a model for one of the best trout lures. This is because their life cycle consists of a long time spent as underwater nymphs, followed by a very short time as flying insects. Huge mating flights of the adults take place above the water, and after only a day or two the adults have laid eggs and died, leaving behind a bonanza of food for freshwater fish.

Mayflies live in freshwater streams and lakes. In temperate regions these water bodies can be very cold, and so development is slow. The larva may take between one and three years to undergo up to 25 moults before emerging from the water. Apart from the very high number of stages, mayflies are also unique in having an extra, short-lived, winged stage before a final moult to a winged mature adult stage. This preadult stage is also well known to fishermen, who call it the 'dun' due to the insects' smoky-grey wings.

Nymphs are flattened, with paired feather-like gills, and biting mouthparts. They live attached to rocks, among underwater leaf litter, or in shallow sandy burrows. Nymphs are mainly filter feeders on minute plant detritus, although some are actually predatory. Because the adults live only for one to a few days, their mouthparts do not develop, as they do not need to feed in this short time. The otherwise useless gut is filled with air to make their flights easier.

In temperate areas, particular species time their emergence exactly, with huge numbers hovering above streams and lakes. These frantic mating flights end after a day or so, during which time they lay from 100 to a staggering 12,000 eggs into the water. In the tropics emergence tends to be more continuous, although it is sometimes tied to the phases of the moon.

The surface of this lake in America is covered in a carpet of dead and dying **mayflies**. These perfectly timed emergence events allow all individuals to have a chance to mate and reinvigorate the gene pool in one short day. For fish it is a feeding frenzy, which is often taken advantage of by fishermen.

A large **mayfly** with a wingspan of 3.5cm (1.4in) from Madagascar. The shimmery rainbow hues in the wings are common to many insect groups, and may confuse a predator as it intermittently flashes or disappears during flight in the sunshine.

A typical **mayfly nymph**. It has the three tails, or cerci, of the adult, but the abdomen has hairy, or sometimes feather-like gills on the sides. The mouthparts are used to scrape algae off the underwater leaf litter, and later are non-functional in the adult.

This small **mayfly** from the Australian Alps is in the brief subimago, or preadult stage. It is characterised by milky-grey wings, which fishermen call a 'dun'. After only hours or a day it will moult one more time and become a full adult. 1cm (0.4in) long.

Jappa sp., in the largest **mayfly** family, the **Leptophlebiidae**. Males often have extra large eyes, which are used to spot females during the frantic mating flight.

DRAGONFLIES and DAMSELFLIES

Order Odonata

6,500 species in 30 families

Dragonflies and damselflies, strong and eye-catching fliers, have long been admired in art and mythology. Apart from some flies, no other insects have such extreme control of flight. The forewings and hindwings can move in separate motions, allowing almost instant changes of direction, even suddenly reversing. All this at speeds up to 40kph (25mph).

Dragonflies were the largest insects known to have existed. Fossils of dragonfly-like insects with wingspans up to 75cm (30in) have been unearthed from the Carboniferous period, around 300 million years ago. Today's widest wingspan is about 25cm. Insects have no lung equivalent. They breathe through a sequence of little holes, called trachea, along the sides of their bodies. As there is no pump, getting oxygen to everywhere it needs to go inside the exoskeleton body is one of the limitations to their size. However, the Carboniferous period was so plant-rich that it had higher oxygen levels in the atmosphere, which helped the efficiency of this breathing system.

Most species have fully aquatic nymphs, with gills and a hunting lifestyle. They roam around or wait in ambush for a variety of prey, which includes everything from other insects to worms, tadpoles and fish. The types of water utilised vary from mud pools, puddles and tree holes, to lakes, streams and rivers. Development time is tied to temperature and food. In the tropics some puddle-dwelling species can go from egg to adult in just 30–40 days, while some of the bigger species in cold climates may take up to six years! After 9–15 moults, they emerge onto waterside vegetation and moult into the winged adult.

This order is divided into the two distinct groups. The main differences between damselflies and dragonflies are discussed below.

Anisoptera — the Dragonflies

Dragonflies are more robust than damsels, with fatter bodies, large heads, and eyes often big enough to meet in the middle. When at rest their wings are more often spread out flat, although they can be together. Their aquatic nymphs are also more squat, and their gills are mostly hidden in a cavity at the rear end. The nymph heads are remarkable for having an adaptation known as the 'mask'. Their mouthparts have fused into a long, multi-jointed structure which folds under the body at rest, but can shoot forward into a long straight extension of the jaws. Many are big enough to use this rapid surprise method on vertebrate prey such as tadpoles and fish.

This surprisingly magenta-bodied male **dragonfly** guards sunny ponds, where it can best show off to visiting females. It is found throughout South-East Asia. *Trithemis aurora*, **Libellulidae**. Wingspan 6cm (2.4in).

Shaded wings are more common among **dragonflies** than mayflies. This male **Iridescent Flutterer**, *Rhyothemis braganza*, from tropical Australia is displaying to potential mates.

A rare flight shot of a dragonfly. Their fast darting flight patterns are designed to escape predators such as birds, although at this Australian swamp some birds did get a feed. Note the streamlined position of the legs in flight.

Red is common in **dragonflies**, although rarely adorning the extremely intricate vein system of the wings. These large wings can beat as independent pairs, providing the control for the best aerial acrobatics in the insect world. This male *Neurothema* sp. is from New Guinea. Wingspan 7cm (2.8in).

Dragonfly eyes are not only large, but composed of a very large number of separate facets. Each one of these tiny divisions is an eye in itself, with a lens and retina, and receives a separate image. They can see in every direction at once. Although they cannot change focus, any movement passing across more than one facet is detected more acutely than a single eye can – a perfect system for hunting. Especially as some dragonflies can have up to 28,000 separate facets.

The highlands of New Guinea's Bougainville Island, lives this very showy **dragonfly**. Like most species it perches on the tips of fine branches or even just grass, and surveys its hunting and mating territory. Look very closely on the clear part of the hindwing and a microscopic fly comes to view – it is a biting midge, same group as the infamous Highland midges of Scotland, here taking blood from the dragonfly. *Rhyothemis resplendens*, **Libellulidae**. Wingspan 6cm (2.4in).

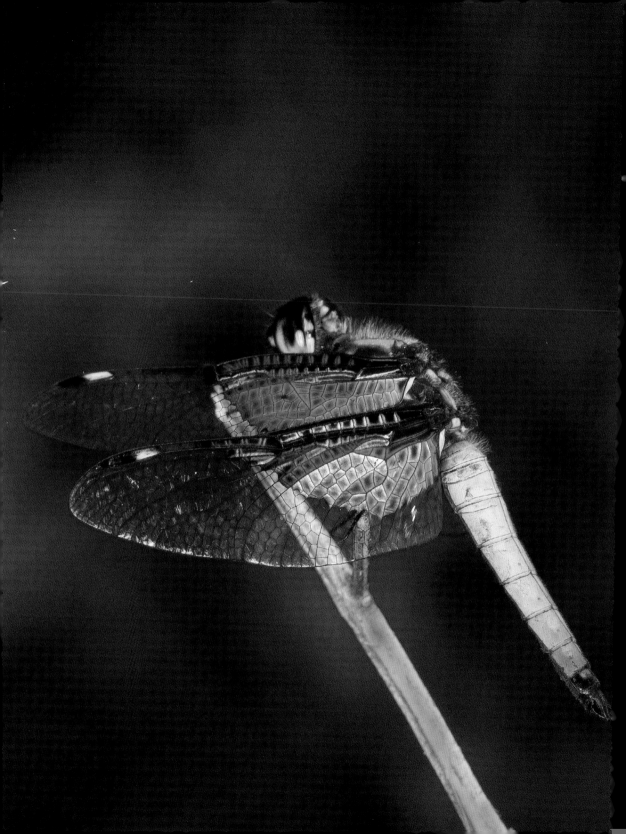

Fire-engine red has nothing on this severe red **dragonfly** from the highlands of Madagascar. It breeds in shallow pools formed during the short wet season.

Dragonfly nymphs are more robust than damselfly nymphs. Also their gills are mostly hidden in a chamber at the back, with just three little triangles protruding. Some are huge, up to 6cm (2.4in) long, capable of hunting fish larger than themselves.

One of Europe's grandest **dragonflies** is the **Broad-bodied Chaser**, *Libellula depressa*, from the family **Libellulidae**. This is an adult freshly emerged from its last nymph, and is still drying in the sun. Later, the blues and yellows will deepen even more.

A female **Common Darter dragonfly**, *Sympetrum striolatum*, of Europe, belonging to the family **Libellulidae**. Males are dark red.

Opposite: One of the most stunning of the reflective shades found in **dragonfly** wings belongs to this amethyst blue species from Madagascar's interior, *Palpopleura vestita*. Here the wings are in a downward pose, which damselflies never adopt.

Zygoptera — the Damselfies

Damselflies, as the name implies, are the more dainty group. Their bodies are generally smaller, more slender, and the wings narrower and of similar size, normally held upright at rest. Their aquatic nymphs have three feather-like gills at the rear end. Unlike the fiercely territorial dragonflies, most adult damsels leave the water area and hunt in habitats nearby, returning only to mate and lay eggs. Mating rituals can be quite complex, and the mating position like no other. The male clasps an area just behind the head of the female. They can then be flying around either in a 'wheel' position while mating, or with the male upright above the female, using his flight to hold the female just above the water as she lays eggs. Damselflies are predators as nymphs and as adults. Small insects such as mosquitoes are often devoured, and sometimes other damselflies also fall prey.

This is not the most beautiful **damselfly** in the world, but it is the largest – *Megaloprepus coerulatus*. It has the widest wingspan of any living damselfly or even dragonfly, at up to 19cm (7.5in). It flies slowly with ponderous wingbeats which flash black, blue and white. An amazing sight in the rainforests of Central and South America, they are known as 'helicopter damsels'.

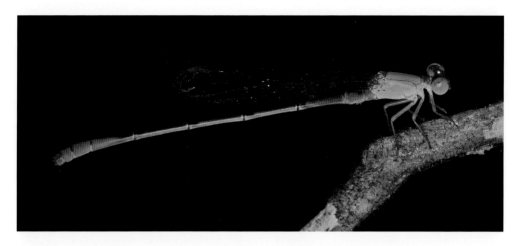

Ceriagrion cerinorubellum from Sri Lanka – a blast of brightness in the rainforest gloom. 3.5cm (1.4in) long.

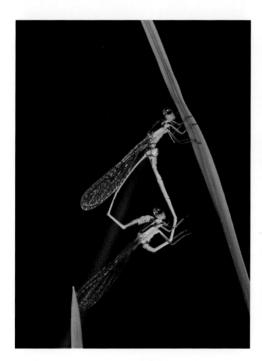

Damselflies have a unique mating position, with the male clasping the female just behind the head. This tandem positon is replaced with the 'wheel' position when actual mating occurs, here shown by a species from the Australian tropics. Later the male still holds the female and hovers above the water, holding her aloft as she lays eggs.

One of the most species-rich genera of **damselflies** is *Ceriagrion*, which is widespread from Australasia to South-East Asia. These small damsels look innocuous, but are not above cannibalism. Here a northern Australian species is eating another damsel of similar size.

This *Rhinocypha* sp. of **damselfly** is a delight along rainforest streams in Indonesia. The metallic rainbow effect is caused by pits in the wing surface that reflect light from different angles. As it flies its wings flash or just appear black, which makes it intermittently invisible to predators. 3.5cm (1.6in) in length.

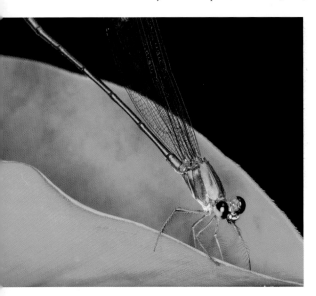

Metallic species are common in many insect groups, but less so among **damselflies**, making this species, *Vestalis amethystina* from New Guinea, very special. 4.5cm (1.8in) long.

The proportions of **damselflies** are varied, from squat species where the body is shorter than the wings, to oddly long-bodied species where the body is up to several times the length of the wings. This species from Indonesia is not the longest, but its abdomen is long enough to hang during flight, slowing it down compared to more compact damselflies.

This handsome **damselfly**, *Metacrina miniata*, lives near mountain streams in Central America. Body 2.5cm in length.

Above: The **damselfly** *Rhinagrion elopurae*,
Megapodagrionidae, watching over its small patch of
rainforest creek in Borneo. 3cm (1.2in) in length.

Right: Exotic insects are often from the tropics –
except for the wonderful exceptions to the rule. This
metallic **Banded Demoiselle**, *Calypteryx splendens*
(**Calypterygidae**) is from Poland. 3.5cm (1.4in) long.

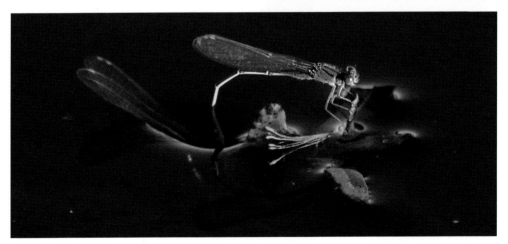

Red-and-blue Damselflies, *Xanthagrion erythroneurum* (**Coenagrionidae**) from Australia in one of two mating poses,
called the wheel. When they are ready to lay eggs, the male only clasps the female on top of the thorax and they fly above
the water with the female dropping eggs just under the surface. But in extreme cases they both descend underwater and
lay eggs in safer locations. Here the male has just started pushing the female under, not yet breaking the surface tension.

STONEFLIES

Order Plecoptera

2,300 species in 15 families

Stoneflies are a small group of insects with an aquatic nymph stage, and mostly winged adult stage. Though unrelated to the mayflies, there are some similarities. They have two tails, called cerci, unlike the three of mayflies. They are also short lived as adults, and most are herbivorous, with only a few families hunting as nymphs. Like mayflies they need clean oxygenated water, living mainly in streams and lakes. Their development depends on temperature and food, with some northern hemisphere species taking several years to undergo up to 33 moults as nymphs, before emergence. Stoneflies hold a bit of a record for aquatic insects, with species living in the Himalayas at altitudes of up to 5,600m (18,300ft) in very cold water indeed. A few species never leave the water, including one which lives at the bottom of a lake In America, at depth of 60m (200ft).

Australia is the driest continent on the planet, with drought a fairly ordinary state. Some **stonefly** species of the genus *Dinotoperla*, in the family **Griptopterygidae**, have evolved very tough eggs that can wait up to 18 months for a dry spell to abate.

Adults live a short life, from a few days to a month or so, and feed on algae, bark and lichens. Emerging adults fly or crawl to high places and mate with little ritual. During this time they can lay from a few hundred to up to a thousand eggs to restart the long aquatic cycle. Females lay eggs either in jell attached under leaves over water, or into the water, or crawl underwater and attach them to stones. While laying eggs, stoneflies repeatedly come close to water, and fish, especially salmon and trout, await this opportunity very eagerly. Fishermen know that if they time their visit to a stream during an emergence of stoneflies, and use similar casting lures, their catches will be memorable.

In North America fly fishermen are very aware of stoneflies. One of the giants of the stonefly world, the '**salmonfly**', genus *Pteronarcys*, reaches 7cm (2.5in) in length. Their flights over water drive fish into a feeding frenzy that can be taken advantage of by fishermen imitating this short-lived insect.

The nymph of the giant **salmonfly** stonefly is larger than most aquatic insects, at up to 6cm (2.5in) long. It spends up to three years growing through 20 or more moults, eating algae scraped off rocks. The adult is a favourite model for fishermen's 'fly' lures.

This bright yellow **stonefly**, more noticeable than its brownish cousins, lives in the cloud forests of Costa Rica. Here, at altitudes of 2,500m (8,200ft), moss-filtered streams are cold and clean, and full of fine algal food for its herbivorous aquatic nymphs. Although in the tropics, where stoneflies often have two or more generations in one year, this species takes two or more years to develop in the cold at high altitude.

COCKROACHES

Order BLATTODEA

4,500 species in 6 families

It is unfortunate that a few species of this diverse group have been spread around the world, and that their reputations as pests gives the entire family a bad name. The other 4,500 species live their lives in natural habitats, contribute to the recycling of nutrients, and sometimes even live in cooperative families. Cockroaches have been around since before the dinosaurs. Most species are nocturnal, and eat mainly living and dead plant matter. Some very bright species have adapted to diurnal lives, even in deserts, where other species would dessicate.

Little is known of cockroach courtship, although in some species males not only produce pheromone 'smells' to attract the females, but these are edible liquids to reward her attention. Females create a leathery pouch around clutches of eggs, and then glue these pouches in suitable areas. Some develop the eggs internally, and some live communal lives and care for their offspring inside complex tunnels.

Many cockroaches produce defensive fluids. The milky mush released when an American Cockroach, *Periplaneta americana,* is squashed is an example. These are not very toxic, but serve to ward off predators whose sensory organs may be clogged with the sticky fluid. The American Cockroach, the red scourge of urban dwellings, is also one of the fastest-running insects in the world. They have been clocked at 5.5kph (3.4mph), which is a brisk walking speed for a human, but the size difference makes this a staggering figure. Measured in body lengths per second (bl/s), the roach covers 50 per second. The fastest-running human covers only 6bl/s! The acceleration of the roach is many times the rate of any vertebrate animal, or sports car, or what an astronaut could survive.

The most interesting species are those most people never see. Many species in two families, especially the **Nocticolidae**, have adapted to caves, with totally blind and even eyeless species still being discovered in cave and lava tube systems all over the world.

Opposite: One of the giants of the cockroach world is Australia's **Giant Burrowing Cockroach**, *Macropanesthia rhinoceros*, which grows up to 8cm (3.2in) long and weighs up to 30g (1oz). There are mammals, such as small bats, which weigh only 5g (0.2oz)! This monster is a gentle creature, living in complex underground societies of many individuals, each taking care of their young. It makes hissing sounds when disturbed and is a popular pet in Australia.

Right: Although not as large as the Australian roach, the **hissing cockroaches** of Madagascar have similar habits. Still huge at up to 7cm (2.8in), they live in groups inside rotting wood, which they can digest with the help of bacteria similar to those found in termite guts. They also care for their young, one of which can be seen at the bottom of the picture. And yes, they hiss, producing the sound by releasing air from their breathing holes, the spiracles. Four different sounds can mean anything from males about to fight, to nest intruder alarm. This species is *Gromphadorhina portentosa*.

One of the smallest roaches, at about 0.5cm (0.2in) long, is *Nocticola*. This delicate, almost see-through, **blind cockroach** lives in lava tubes and caves in northern Australia. It lacks body pigment and eyes and spends its entire life in total darkness.

The largest cockroach in South America is *Blaberus giganteus,* in the **Blaberidae** family. Up to 7.5cm (3in) long, it also lives in groups, often in extensive wood tunnels dug by other animals before them. These spaces are often shared with bats, whose guano is a major source of food for these omnivorous roaches. Being active mainly at night, and good at rapidly digging, helps to protect them from the most serious roach enemy, the army ants.

There is a behavioural story here which is both bizarre and very scary. There are wasp parasites of cockroaches which want living meat for their babies to eat. Most parasites just paralyse their prey. But this one, *Ampulex* sp., the jewel or **emerald cockroach wasp**, injects a neurotoxin which turns the roach into a zombie. It can and does move, but only where the wasp coaxes it to, but has no free will to escape. It is thus led to a lair in which the wasp lays an egg. When the larva hatches, it devours the roach slowly until ready to pupate into an adult that starts the macabre sequence again.

Not all **cockroaches** are dull browns and blacks. In Central America live species of *Euphyllodromia*, which are considered to be mimics of local wasps. Apart from the strong patterns, they move in wasp-like bursts of frantic walks after alighting, and are very quick to fly off.

Left is a nymph of *Ellipsidion australe*. The striking pointillist nature of the markings bear a resemblance to the art style of many Australian Aboriginal people. Below, the golden adult of this species is wildly different from its immature stage.

The green **banana cockroach**, *Panchlora* sp., stands out from normal brown species and is found from the southern USA to the Caribbean.

Cockroaches lay their eggs into leather-like bags called oothecas. These are often carried about for safety, and later glued to a surface such as bark. *Cosmozosteria gloriosa* is shown here.

The Australian arid zone, which covers most of the continent really, is home to remarkable day-active **cockroaches**. Many of them are amazing metallic species. Some are involved in pollination, others in general nutrient cycling by eating plant detritus, but all are the opposite of the image and habits of domestic pests.

Polyzosteria mitchelli, **Blattidae,** 2.5cm (1in).

Desmozosteria elongata, **Blattidae.**

Balta insignis, **Ectobiidae,** 2cm (0.8in).

Cosmozosteria maculimarginata.

TERMITES

Order BLATTODEA (Isoptera)
2,700 species in 6 families

Termites, like the cockroaches they are related to, are much maligned. As with the roaches, the majority of species live interesting, useful lives, away from bothering humans and their structures.

These ingenious architects of the most complex structures in nature are the major recycling force on the planet. Most species live away from our homes, in nests that can house literally millions of termites. These mounds and subterranean structures are air conditioned to within 2°C (3.6°F), even in extreme habitats, and are built and guarded by insects so well adapted to their tasks that it is no impediment that they are blind. The largest nests recorded are up to 10m (32ft) high and 15m (50ft) wide at ground level, and new evidence shows that some have been inhabited for more than 1,000 years. Termite lives are very short. Workers and soldiers may only last a few weeks, although their huge queen can live for 20 years and in some cases up to 50 years.

The life cycle of this social insect starts with the production of huge numbers of winged termites of both genders, called alates. Once a year, when the weather conditions are just right, they leave the nest and join other flights of the same species. They are very vulnerable as their soft, fat-laden bodies are a great favourite fattening food source for many small mammals, birds, bats, frogs and insects. As soon as they have mated the couple drops to the ground, discards their wings and starts digging a new nest. The female becomes the queen of this new colony, although only a tiny fraction of the couples survive long enough to make a new nest. As her first eggs hatch, she has workers, and then later soldiers, to look after her and keep building the nest. The queen keeps growing into a huge distended egg-laying machine, fed and cleaned by a horde of workers. She lays hundreds of eggs each day, and up to 30,000 per day in large mound species. The largest queen found, in the genus *Macrotermes*, was nearly 15cm (5.9in) long, dwarfing the workers which rarely exceed 0.5–0.6cm (0.2–0.24in).

The main food of termites is plant matter, more often already dead and attacked by fungus. This makes it easier to digest, although termites are the only creatures to possess a single-celled gut protozoa, which can actually break down cellulose. In some areas, such as semidesert grassland, the plant matter is not attacked by fungus in the dry atmosphere. Species of *Macrotermes* have evolved to farm a fungus species inside their nests, and use this as their main food. They chew up plant matter

In the tropical north of Australia lives a termite – *Nasutitermes triodiae*, in the family **Termitidae** – that builds different style nests in different habitats. The very tall towers, measuring up to 6m (20ft), are found in grassland. Exposed to the sun, the tower is open at the top, releasing hot air while sucking in cool air from underground. In partly shady woodland, the so called 'elephant's bum' nest design takes over. Both nests can hold many hundreds of thousands of these harmless grass-eating termites.

and mix it with their faeces. This is what the fungus grows on, in elaborate humidity-controlled gardening chambers.

Termites are very soft creatures, and generally cannot survive outside in daylight in low humidity. Therefore most are active only at night, or inside the highways of mud tunnels some species use to get from the nest to food sources. In tropical rainforests though, the humidity is very high, and so several species forage in the day. Some even move away from

nests, transporting the very well-guarded queen in the open.

Soldiers can be equipped with a variety of jaws, from short strong triangles, to long scimitars. The most interesting are the 'nasute' termites, a group of species which has evolved a snout, which fires a foul, glue-like substance at attackers. Ants are the main enemy of termites, and this method works well, as fastidious ants cannot tolerate having their antennae and mouthparts covered in muck.

And the pests? Some species which eat wood as their main plant matter have adapted to invade our timber homes. In the wild they would have lived on the heartwood of living trees, or any fallen trees, but a house-sized structure of 'fallen' timber is a huge attraction. Despite the very real damage they cause, it is important to realise that they are a small fraction of species within this fascinating group.

Early explorers of north Australia were intrigued by these tall, very flat nests, all lined up in a perfectly north-south orientation. These are the **'magnetic' termites**, *Amitermes meridionalis*, in the family **Termitidae**. We now know that they have a working compass in their heads, based on minute amounts of iron. The design of the nest is to warm up at sunrise, but then present a tiny thin wedge to the sun at blisteringly hot noon. These 3m (10ft) tall structures are often whitish, as the termites tend to live on kaolin-rich clay soils.

A very purposeful army of termites marching along the floor of a rainforest in Borneo. This species, *Longipeditermes longipes*, in the family **Termitidae**, can be active day and night in the high humidity here, and forages for a variety of mainly dead plant matter.

The soldier caste come in all shapes and sizes. From minute 0.2cm (0.08in) ones to the largest at 2.2cm (0.9in). The variety is in their jaw structure, from short spades to long 'swords' to the bizarre turrets which spray toxic glue onto enemies. A selection is shown below.

Macrotermes carbonarius, **Termitidae**, Thailand, 1.2cm (0.5in).

Neotermes insularis, **Kalotermitidae**, Australia, 1.2cm (0.5in).

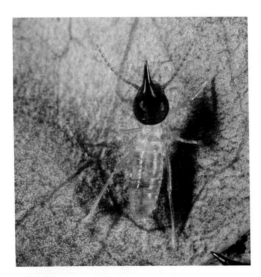

The **nasute termite**, with turret, *Nasutitermes triodiae*, **Termitidae**, Australia, 0.6cm (0.25in).

Hapsidotermes maideni, **Termitidae**, Australia, 0.4cm (0.15in).

Some soldier termites look quite formidable, but they are still a softer and usually smaller opponent than their arch enemy, the ants. In a one-on-one battle with the faster, tougher ants, the result is not pretty. Termites have to use sheer numbers to try to overpower and push back an attack. Left an *Iridomyrmex* ant dominating in the battle, and right, the famous Matabele ant of East Africa, *Megaponera analis*, carrying away a mouthful of defeated termite soldiers.

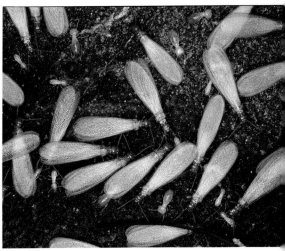

The monstrously distended queen termite. The queen spends her long life, up to 50 years in some species, inside the royal chamber, in the dark, attended by hordes of workers. She is joined by the king, the only male in the colony (the large armoured termite facing her head). Here she lays up to 30,000 eggs per day, which are taken away by the pale nursery workers who also take care of the huge growing brood. *Apicotermes* sp., **Termitidae**, from Mozambique.

The alate caste are the winged reproductives, produced and released once a year to start new nests. Here a *Neotermes* species from the family **Kalotermitidae** are milling outside the nest in a tropical forest in Australia, fussed over by tiny workers and protected by soldiers. The alates are the biggest caste, as their bodies must have the stored energy to fly away, mate, and actually build and start a new colony.

MANTIDS

Order MANTODEA

2,200 species in 8 families

The praying mantis is a very finely adapted predator, from a very mobile head with large keen eyesight, to the rapid striking force of their forelegs equipped with rows of spines. Their often cryptic appearance, and the very disciplined stillness of an ambush predator, add to the list. Many mantids are well camouflaged by a stick-like, bark-like or flower-like body. They sit still and wait for an insect to get within range, only very slowly moving their heads to keep a distance-measuring stereo image, before pouncing at high speed. The foreleg spines partly crush the prey, and hold fast.

Like black widow spiders, they are also famous for very dangerous females. Although not normal at every mating, it is true that the females will eat the males during mating, given a chance. The rationale is simple. Get an instant feed, enough to produce many already fertilised eggs. The males are smaller, and approach her very, very carefully from behind. If they manage to jump squarely onto her back, she cannot reach with her spiny forelegs, and mating will occur. During this time the female is in a slowed-down state, but as soon as mating is completed she is fast again, and the male must jump off and run. This cannibalism could be disadvantageous to the species, if males are eaten before they can mate. Mantids are solitary, so not all get to meet a prospective mate. However another adaptation has evolved from this behaviour. Males actually mate with greater vigour when their heads are off! The female always starts by eating the head, and this switches on one of the ganglia (small

'sun-brains' in other parts of insect bodies) to devote more effort to mating.

Eggs are encased in a leathery, and then froth covered structure called an ootheca. She often attaches this to branches, with anywhere from about 10 to 400 eggs.

The baby mantids are wingless, skinny copies of the adults. They uncurl their elongate bodies out of the eggs, and run away from each other starting their hunts. Adult sizes vary from about 1–16cm (0.4–6.5in).

The majority of species are tropical. Some very ornate cryptic bodies, and fabulous flower mimicking bodies are found, especially in Asia and Africa. In Australia there are stout little wingless mantids in the family **Amorphoscelidae**, which often mimic ants when young. Other mantids also sometimes have a first-stage nymph which imitates ants. This adds protection during their early stages, as ants are typically disregarded by predators as too much trouble. Most species are in the

A typical large tropical **mantis** from Australia, *Hierodula* sp. What is both most endearing and most creepy is the way their heads follow your every movement. They may be very still otherwise, but their ambush hunting skills are always hyper-engaged, looking straight back at you.

This blue-eyed beauty demonstrates one of the features of **mantids** that make their interest in us so pronounced. The little black dot in each eye looks like the pupil, and makes their stare all the more personal. It is in fact only a dot of pigment on the outside of the compound eye, which is composed of hundreds of individual facet eyes.

more typical family, **Mantidae**, with adults more often winged.

The most surprising fact about mantids however is to do with their hearing. Almost all creatures on the planet have sets of two (or more) sensory organs such as eyes, ears, nostrils, etc, so as to determine the direction a sight, sound or smell is coming from. Mantids however have just one ear, a 'hole' between the last pair of legs. It cannot detect direction of sound, but recent studies found that it hears only high-pitched sounds such as those of bats. It is therefore a warning organ for oncoming bats when the male mantids fly at night. They don't need to know the exact location of a bat, just that it is time to fold the wings and dive to the ground to get away from it.

Flower mimicking is an adaptation evolved among many species. It adds even more sneakiness to the ambush, to look like part of the flower an insect wants to land on. This species from the genus *Pseudocreobotra* is from West Africa. 2cm (0.8in).

One of the weirder members of the **mantis** clan, is this super thin species from Borneo. Limited in strength by its fragile structure, it hunts lighter prey, such as midges, mosquitoes and the like.

Mantids protect their eggs in a froth which hardens to the consistency of packing foam. Here a mantis in Borneo is finishing one of these formations, which is called an ootheca.

One of the most extreme flower adaptations are among the **orchid mantids** of tropical Asia. This individual from Malaysia, *Hymenopus coronatus*, is sitting on a flowerless stalk, and is not just trying to blend with an orchid, it is the orchid.

Two delicate **flower mantids**. Top: *Junodia amoena*; and above: *Pseudocreobotra wahlbergi* from Mozambique. Even when not among other flowers, the convoluted flower-like bodies of these mantids may attract flower-loving insects to their doom.

The most iconic African mantids belong to the families **Empusidae** and **Sibyllidae**, and are known as the gargoyle or **cone-headed mantids**. Top are late instar nymphs of a *Sibylla* species from West Africa, and above, the only member of this genus in East Africa, is the **Precious Sibyl**, *Sibylla pretiosa*, in Mozambique.

Praying mantids tend to be sluggish flyers, so a first response to threat is not always to fly off with gusto, as a fly would. But their forewings often hide a surprise set of warning markings on the inside, with extra warning patterns on the hidden rear wings. The large *Omomantis zebrata* from Mozambique, suddenly unfurls its wings as a startling display to slow down the response of a predator. Body about 5cm (2in) long.

Although stick insects are usually talked about as camouflage masters, mantids have some hiding strategies that defy even their wiles. It is partly to get closer to prey, but also to be invisible to their own predators such as lizards, birds and mammals.

Australia is a very dry and hot continent, where bush fires are common and many plants need fire to regenerate. Therefore the trunks of eucalypt trees are often blackened, and a species of **mantis** in the genus *Paraoxypilus* has adapted to slowly darken its body to blend in perfectly.

Namibia is a country of spectacular deserts, savannas and canyons. Tough grasses spend much of their lives in dry dormant states. Among these yellowing stalks hides a perfect ambush predator, the **grass mantis**, *Epioscopomantis* sp.

In the rainforests of South-East Asia the leaves are usually large, and the leaf litter forms a dense habitat for a myriad of creatures. It is a place where mimicking dead leaves is advantageous for both predators and prey. This species from the genus *Deroplatys* is typical of **leaf mantids** in Borneo.

In the wet cloud forests of Costa Rica not all is green. A bunch of dead leaves on broken branches is a habitat in itself. Moths, bugs, stick insects and even geckoes hide here. One amazing **mantis** in the genus *Acanthops* not only mimics the leaves themselves, but breaks up its body shape by fully bending in the middle. Here it awaits other insects looking for shelter, while remaining invisible to its enemies.

When insects use crypsis to hide against backgrounds of a similar pattern and texture, it is often only some aspect of symmetry that gives them away. Here the body of *Otomantis scutigera* from Mozambique is perfectly blended with the bark, but the symmetry of the eyes gives it away to our perception. However, most predators hunt by searching for whole shapes and/or movement, making this disguise very effective protection.

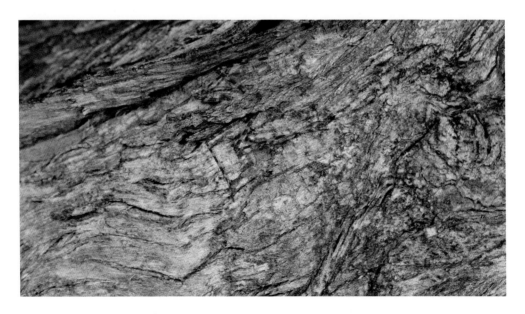

This **mantis** from Western Australian arid country is a perfect copy of the bark of this eucalypt tree. 2.5cm (1in) long. Good luck with finding it.

HEEL-WALKERS, GLADIATORS, ROCK-CRAWLERS

Order MANTOPHASMATODEA (NOTOPTERA)

12 species in 3 families

This tiny order is fascinating for its attributes and even more so for its very recent discovery story. In 2001 a German entomologist published a paper about two insects found in 45-million-year-old Baltic amber, which simply did not fit any current order-level description. Added to the discovery of two more museum specimens of recent age, this new order was formed.

What was so alarming about these insects was their mix of traits from crickets, stick insects and mantids. They are wingless, cricket-like insects with mantis features and carnivorous habits, and oddly bent 'feet', whereupon they walk not on the tips of their tarsi, but more like the 'heel'.

An expedition in 2002 searched the Namibian habitats of the museum specimens and discovered the first living examples. The photos below were taken by a member of that expedition, Piotr Naskrecki. After this, discoveries in north-western South Africa and in Tanzania followed. In fact some species in Namaqualand have proved to be locally common but unnoticed, as their resemblance to immature katydids made them poor candidates for taxonomic studies — it is hard to identify and impossible to describe a species without an adult.

Heel-walkers are about 1–2cm (0.4–0.8in) long, wingless, with eyes reminiscent of mantids, multi-segmented antennae, and more or less equal-length legs. They tend to live on shrubs in dry habitats, and hunt other insects. Males are much smaller than females. Mating habits can include several days of contact, and as with mantids, the female will sometimes eat the male to feed her egg production. The eggs in the studied species are encased in a foam-like packet and buried in sandy soils near the surface. This is similar to grasshoppers, which encase eggs in large foam parcels deep in soil, and mantids which make elaborate foam homes (an ootheca) for their eggs, which are glued to vegetation.

The wonderfully named *Tyrannophasma gladiator* was the first living **heel-walker** to be found after the order was formed in 2002. It is from the beautiful Brandberg Plateau in Namibia. Right is a South African species of *Sclerophasma*.

ICE-CRAWLERS

Order GRYLLOBLATTODEA (NOTOPTERA)

28 species in 3 families

This is a very small order of insects, with only 28 named species which have very distinct habits. The name suggests a mixed relationship to other insect groups. They have features of crickets, cockroaches and earwigs, although they are closest to, but still distinct from, the crickets. All are wingless, elongate and between about 2–3.5cm (0.8–1.4in) long. The name ice-crawlers is apt, as they do indeed hunt on the edge of ice fields in the spring, and live in the thin open space between the soil and snow through the winter. Unlike some other cold-adapted organisms, they do not possess any 'antifreeze'-like chemicals in their blood, and so will die if exposed to temperatures below about -8°C (17°F). However their winter space is insulated by the snow layer above, and remains around 0°C (32°F) most of the time. The same goes for the ice caves that some of them inhabit. This is a contrast to their ancient lineage, which had hundreds of species in tropical forests as far back as 250 million years ago. Competition with other insect groups drove them to their narrow habitat, and as the last ice age retreated, they followed the ice fields north.

Being cold-blooded and living in cold habitats means that their development is slow. Females lay from about 30–150 large black eggs which take up to a year to hatch. The seven stages of nymphs can take seven years to reach the adult stage. Food is obtained mainly through hunting other insects, although dead insects and some plant matter are taken during the winter.

About half the species are from montane habitats in northern North America, and the remainder are from northern Asia, including Japan and Korea. The Korean species are cave dwellers.

A Canadian **ice-crawler**, *Grylloblatta campodeiformis*, foraging on ice.

The Asian **ice-crawlers** are not all restricted to cold mountains, and can tolerate higher maximum temperatures. Some are cave dwellers or live deep in the soil, where, like many dark dwellers, *Galloisiana nipponensis* forages without eyes.

EARWIGS

Order DERMAPTERA

1,800 species in 9 families

This small order of insects attracts attention more for their name, and the mythology attached to it, than anything else. Gardeners know them as mild pests, as the diet of many species is omnivorous. The Common Earwig, *Forficula auricularia*, is a European species that has been transported to the Americas and other parts of the world. It likes gardens as they provide many crevices in wooden and stone structures, where their bodies can hide during the day and in winter. Their food consists of living and dead plant matter, but also of small soft insects such as aphids and plant lice. As these can be more injurious to the garden than earwigs, it is a mixed bag of good and bad. It is the habit of eating pieces of soft fruit, such as strawberries, that makes gardeners most upset. However this earwig is only one species of more than 1,800, and most would never be found in a garden. Some have similar diets in the wild, others are more predatory, or eat fungus. The oddest families are those which have evolved ectoparasitism, living in the fur of bats, sometimes in the bat guano in caves, and some only on giant rats in Africa. These species lack the pincers, and give birth to live young – a habit of many parasites, to be sure that the young have a host available. Earwigs range in size from tiny soft parasitic species at 0.5cm (0.2in), to a wingless giant in Australia at up to 6cm (2.5in).

Reproduction is also different in earwigs to most other insects. Apart from a few cockroach species, they are almost the only non-social insects which look after their young. Female earwigs lay a batch of large eggs in a hidden chamber or crevice, and defend them from fungus and predators for the one to two weeks or so they take to hatch. She does not even go out to feed during this time. The young are soft and do not posses the full pincers or wings until after the last moult – usually the sixth – and live for about one year.

Earwigs have a unique body structure, characterised by the pincers at the end, which are very hardened and modified cerci (like the filamentous tails in a mayfly). The forewings are short and leathery covers for the underwings, which have a unique, origami-like fan structure. These flying wings are very thin, and the myriad folds that let them safely squeeze under the tiny covers, require the help of these pincers to fold properly.

Another point about the wings is their superficial resemblance to the shape of an ear.

Here then is one aspect of the 'what's in a name' story. The Latinised, or in this case Greek, origin of the insect order name Dermaptera, literally means 'skin-wing'. This probably refers to the leathery forewing though. The word earwig comes from an old Anglo-Saxon word 'eare-wicga' which does literally mean ear-beetle (or insect). So the persistent myth of earwigs crawling into people's ears has some sort of start here, except that it could equally refer to the shape of the flying wing, as to any habits. So, as they do not invade ear canals as a documented behaviour, the myth is in trouble. And of course earwig is a name in only one language/culture. For instance, in some parts of Japan, the old term is 'chinpo-kiri', which roughly translates to 'penis-cutter'. Now there is a myth story worth exploring.

In the tropics, like many insects, **earwig** species tend to be more showy and extreme. This species from Mexico has pincers almost as long as its body, and this is the female. Males normally have even bigger and more ornate ones. 2.5cm (1in) with pincers.

Another tropical **earwig** species, this time from Thailand, shows bold markings and uses its pincers in a mock aggressive pose to see off a threat. Such displays can save earwigs from an initial pounce by a predator. In some species they are used to capture prey, and in ritual battle between males.

This small **earwig** in Costa Rica is showing one of the omnivorous habits of the group. Flowers are visited more to eat them than to pollinate them, although both actions usually result. 0.6cm (0.25in).

The complex origami-folding wings of **earwigs** are not only a bit ear-shaped, but actually rarely used. Most are fully see-through, but this species has bold patterns, which when unfurled suddenly may startle a predator long enough for the earwig to escape. The pincers are used to help get these complex folds back into the tiny wing cases.

The European **Common Earwig**, *Forficula auricularia*, as seen in many parts of the world, is a dedicated mother. Here she is guarding and cleaning her batch of 40 or so eggs, and later, many second stage nymphs, camouflaged among the soil, are still living with her.

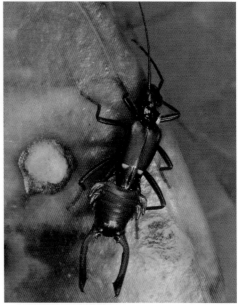

The male **Common Earwig**, *Forficula auricularia*, shows the gender difference in earwigs. Their pincers are larger and more curved than in the female (above). It can put on a show of aggression when cornered, although the most normal function for them is to help fold away the complex wings. 1.5cm (0.6in).

An ornate unidentified **earwig** species from the cloud forests of Costa Rica. 2.4cm (1in) long.

CRICKETS and GRASSHOPPERS

Order ORTHOPTERA

24,000 species in 17 families

This large and very recognisable order is divided into two suborders, loosely, the crickets and the grasshoppers. Both share the basic body plan of large rear legs, usually adapted for jumping. Both have a cover or shield over the thorax, and the top wings are a leathery cover for the fan-like second set of flying wings. Most species can also make complex, loud, sounds by stridulating body parts, and possess ears to hear these.

Wood crickets, Gryllacrididae, remain well hidden during the day. One of their many methods is to utilise hollow branches, especially with a snug fit to prevent other interlopers. This Australian species is a good example of the tight fit. 3cm (1.2in) long.

Suborder ENSIFERA — the crickets and long-horned 'grasshoppers'

The crickets comprise 11 of the 17 families, and have about 60 per cent of all Orthoptera species. Superficially the main difference is that crickets have long, thin, very multi-segmented antennae. The so-called long-horned grasshoppers are in fact part of the katydid cricket family **Tettigoniidae**. Females of some groups have a long curved sword-like ovipositor, a device for laying eggs into crevices. Grasshoppers have four very small tough triangular digging tools instead, as many of them lay their eggs underground.

Crickets are the better, louder singers. The mechanism is via rubbing special hardened veins on the base of the forewings, one like a comb the other more like pegs — a 'tooth-and-comb' technique. Wings can be rubbed slowly for deeper sounds, or at incredible vibration speeds, producing high-pitched sounds way over human hearing range. When describing new species of crickets, having a recording of the song helps to define their uniqueness.

In temperate areas, the main crickets heard at night are the Gryllids, the black classic crickets. In the tropics, the katydids take over the nights with their surprising variety of songs, from simple beeps and chirps, to full bird-like warbling. In daytime most crickets are quiet, and the main noise is that of cicadas. They hold the decibel records for day sounds, with some species quite literally painful to be near. At night, however, the mole crickets are the masters of loud. They live underground in tunnels which often end in two funnel-like entrances. These are very carefully sculpted to amplify sound like the old gramophone horns, and of course invented stereo before we were a species. This sound-throwing ability is so good that an individual can be heard up to 2km (1.2 miles) away.

Crickets are more omnivorous than grass-hoppers. Some feed on living plants, some on detritus, but many are surprisingly adept predators. The large king crickets are known to even capture vertebrate prey such as lizards, frogs or mice.

The most species-rich family is the **Tettigoniidae**, the katydids. They are also the best masters of camouflage in the group, with some very realistic leaf mimicry. This is important in insects which are night active and must evade capture during the day. Other crickets hide in burrows, leaf litter, or inside hollow branches.

The classic field cricket comes in many species all around the world, but varies little from this general form. Black, stout, large headed, and producing a loud song by stridulation of its wings. Shown here is an Australian species from the genus *Teleogryllus* in the family **Gryllidae**. 2.2cm (0.9in) long.

New Zealand is home to famous examples of an island evolutionary trend, whereby some animals, from birds to insects, lose their ability to fly. The kiwis are one, and the giant ground and tree crickets, known as wetas, are another. This **tree weta**, *Hemideina* sp., is 7cm (2.8in) long.

The **Red-headed Bush Cricket**, *Phyllopalus pulchellus*, is one of the more strikingly marked members of the field cricket family **Gryllidae**. From North America.

The **Stenopelmatidae** contains species called **king crickets**, most of which are large stout nocturnal hunters with very fearsome jaws. At more than 6cm (2.5in) long, this species from New Guinea even hunts small vertebrates such as lizards and frogs.

In deserts, most insects and spiders are active only at night. This Australian striped **'wood' cricket**, *Pereremus* sp., family **Gryllacrididae**, sleeps in sandy burrows or under stones or logs during the day, and hunts in the open at night.

The oddest and least-seen crickets are the **'ant-loving' crickets**, family **Myrmecophilidae**. The minute body is cockroach-like and measures less than 0.2cm (0.08in), with very fat rear legs. They spend their entire lives inside ant nests, where they apparently feed off detritus on the ant bodies. Here one is among the larvae of the ferocious bull ants in Australia.

This powerful-looking creature is a **mole cricket**, family **Gryllotalpidae**. They burrow and live underground all day and much of the night, feeding on roots. The very powerful shovel-shaped front legs make short work even of clay soils. But what members of this group are best known for is their very powerful, shrill call, which is amplified via twin trumpet-like openings to the burrow.

Most of the **wetas** are in their own family, the **Anostostomatidae**. The largest species grow to 10cm (4in), but the weirdest are the tusked species. Like many flightless animals, the mainland populations have been decimated by feral predators including rats, but on tiny islands, such as the Mercury group, *Motuweta isolata* lives and hunts. The tusks are only on the males, and used for ritual combat.

The katydids are the largest group of crickets, with more than 6,500 species in the family **Tettigoniidae**. They are responsible for many of the night songs around the world, especially in the tropics. A majority of species are green, and a general leaf-mimicking shape is very common. However this shape is taken to extremes in some tropical species, right down to blemishes, leaf veins and curls, and even chewed holes and edges.

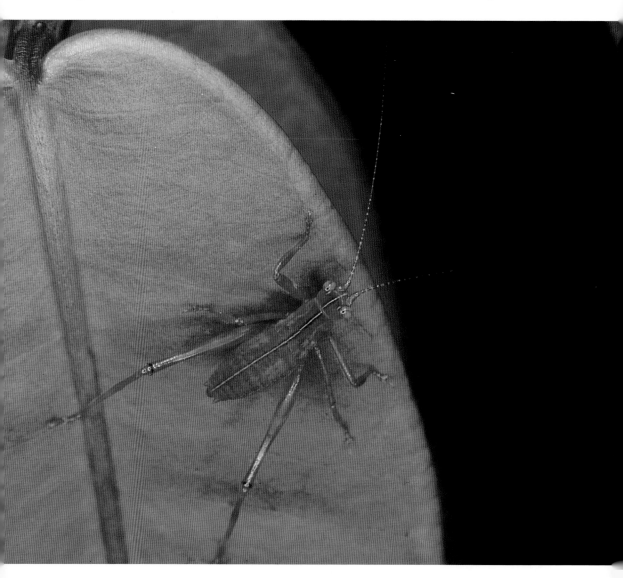

Some insects can slowly change the hue or darkness of their bodies, but a few, such as **katydid nymphs,** can also take on the pigments of their food. This red nymph will later change its diet, and grow up to be a normal leaf-like green katydid.

This eyecatching katydid from New Guinea is a last stage nymph. The short wing-covers will become full wings in the next moult, when it attains adulthood.

All the camouflage in the world cannot help a Costa Rican **katydid** which sits on the wrong backdrop.

Above and top right and left: **Leaf katydids** from Ecuador.

Leaf katydid from Borneo.

Leaf katydid from Borneo, extra well hidden.

Above: **Dead-leaf katydid** from a Costa Rican forest floor.

Right: **Leaf katydid** from Thailand, 4.5cm (1.8in).

One of the most delicate **katydids**, *Acauloplax exigua* from Mozambique. Its very flat body helps it to blend with leaves by casting no edge shadows.

Apart from leaves, katydids also use moss and lichen for disguise. On the right, a **lichen katydid**, and on the left, the very large, 6cm (2.5in) long, almost invisible, **moss katydid**. Both from Australian rainforests.

The **domed katydid** from New Guinea has a different body structure to the usual flattened leaf, and has extra layers of spines to protect its more noticeable form.

Katydids may look like grasshoppers, but they do not eat grass. Here a nymph from Borneo is making a mess eating a moth. 2cm (0.8in).

Katydid nymphs often look very different to the adults. Some species even mimic ants in their very early stages, as this gives them protection from many predators which avoid ants (above). Others just have wilder patterns than their mainly green parents (overleaf).

Suborder CAELIFERA — the grasshoppers

Grasshoppers are a more homogenous group than the crickets. The majority of species are in the one classic family, the **Acrididae**, which has the 'normal' grasshoppers and locusts. Their often cylindrical bodies have leathery forewings covering large fan-like, folded, usually see-through, rear wings. Females do not have an ovipositor, which crickets use for egg-laying. Their mouthparts are very similar in most species too, characterised by a large upper 'lip' called the clypeus, under which are short but very strong and sharp jaws for chewing vegetation. Gardeners know how quickly the lignin in plants dulls the edge of cutting tools. Almost all are vegetarians, and mostly on living plants, from grasses to tree leaves.

The jumping ability of grasshoppers is taken for granted, but it is quite remarkable. Comparing tiny creatures' skills directly to humans is not useful, so one way to get an idea of abilities across the animal world is to measure body lengths (bl). Thus the best of them all, the fleas and certain planthoppers, can jump more than 200bl in one leap, while grasshoppers leap about 30bl, which moves them about 75cm (30in) away from danger. The human long jump record of 8.9m (29ft) is only just under 5bl.

Many grasshoppers have 'songs', but these are not as loud and melodious as the crickets have evolved. The method of producing the sound is different too. While it is still essentially a comb and peg concept, one part of this washboard pattern is on the rear legs, and the other on the upper wings. So a stridulating grasshopper can be seen to move its back legs, while stationary. A singing cricket is more likely to be seen holding up its upper wings while stationary. Grasshopper songs tend to be more like repeating chirps.

A few species in the **Acridiidae** have evolved a complex life cycle in arid areas, which can efficiently take advantage of a rare wet, greening event. They can breed very quickly on the fresh grasses, and when they decimate these, huge numbers then migrate to other areas. These are the plague locusts, with the most famous examples in Africa and Australia.

The second-largest family is the often eyecatching and fat **Pyrgomorphidae**. Many of these feed on plants which are poisonous to vertebrates, and they use very garish bright patterns to warn their prospective predators not to eat them.

Other families include the monkey grasshoppers in the family **Eumastacidae**, and the wonderfully odd stick grasshoppers in the **Proscopiidae** family. Very cryptic and spiny species make up another group, the Pygmy grasshoppers in the family **Tetrigidae**.

The **Pyrgomorphidae** family has many of the extreme species of grasshoppers. Here are a few of the warning patterns used by them to dissuade predators from tasting their poisonous bodies.

The **Leichhardt's Grasshopper**, *Petasida ephippigera*, family **Pyrgomorphidae**, Australia. 4cm (1.6in).

A **pyrgomorph** nymph, **Pyrgomorphidae**, from Madagascar.

Rainbow Desert Grasshopper, *Dactylotum bicolor*, family **Pyrgomorphidae**, from USA. 3cm (1.2in).

Opposite: The **Rainbow Milkweed Grasshopper**, *Phymateus saxosus*, family **Pyrgomorphidae**, from Madagascar, 6cm (2.4in) long.

Above: *Dictyophorus spumans*, **Pyrgomorphidae**, from South Africa.

Right: The **Common Milkweed Grasshopper**, *Phymateus morbillosus*, family **Pyrgomorphidae**, from the South African highlands, 6cm (2.4in) long.

Taphronota ferruginea, **Pyrgomorphidae**, from Ghana.

Wingless **pyrgomorph** at 4,000m (13,000ft) on Mt Kilimanjaro, Kenya.

The **Eumastacidae** family has a subfamily, the **Biroellinae**, known as the **monkey grasshoppers**. Most have short bodies with extra-long rear legs which they hold out at a 90-degree angle to the body. This stance combined with often wild patterns makes them quite entertaining.

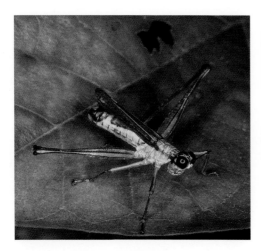

Yellow Monkey Grasshopper, *Paramastax* sp., family **Eumastacidae**, from Ecuador. 1.5cm (0.6in).

Capy York Monkey Grasshopper, *Biroella* sp., family **Eumastacidae**, from Australia. 2cm (0.8in) long.

Above: **Monkey grasshopper** from Madagascar in full 'cross' pose.

Left: Some **monkey grasshoppers** do not hold their rear legs flat, but still stand out from the crowd with wild patterns, and the large and complex-looking last segment of the abdomen. Here a species from Indonesia. 1.5cm (0.6in).

The **locusts** are a serious scourge that originates in arid areas of Africa, Australia, the Middle East and South America. They breed in massive numbers after rare rains green the deserts, and then migrate to better food sources as giant clouds of hungry marauders. In a new location they again eat all vegetation available and, weather permitting, repeat the breeding and migration cycle with even bigger numbers.

The **Australian Plague Locust**, *Chortoicetes terminifera*, family **Acrididae**. At about 3cm (1.2in) it is a smaller grasshopper than the classic Desert Locust in Africa, but can be equally destructive. Here first generation females are laying about 40 eggs each into deep holes which are then sealed by a special froth. The huge next generation of wingless nymphs hatches and marches across the countryside eating most plants on the way.

The large 5cm (2in) long African **Desert Locust**, *Schistocerca gregaria*, family **Acrididae**, can eat its own body weight of vegetation every day. Migrating swarms can cover several square kilometres, and a single area of 1 sq km (0.4 sq miles) has enough locusts to eat a total vegetation mass that could feed 35,000 humans. Above is a flying swarm in Madagascar, maintaining this density as far as the eye can see in every direction.

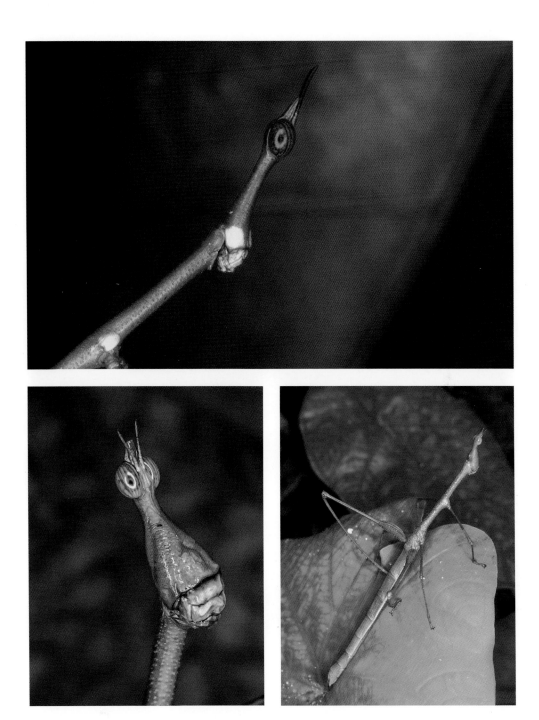

Amazonian South America, home to a riot of insect species, is the key area for the **stick grasshopper** family, the **Proscopiidae**. At first glance they appear very similar to stick insects, but a closer look reveals bizarre faces like no others.

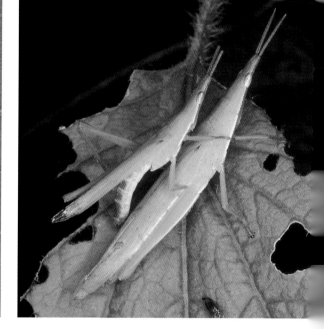

Top: *Diabolocatantops axillaris* nymph looks nothing like its 'normal' grasshopper adult, but helps it blend with amaranth flowers. From Kenya, 2cm (0.8in).

Above: *Valanga irregularis* is a large **grasshopper** found in South-East Asia and Australia. Here second and fourth instar nymphs are sharing a meal. 1–6cm (0.4–2.4in).

Top: The **Zebra Grasshopper**, *Zebratula flavonigera*, is one of the signals of the start of the wet season in Australia's Top End. 2.5cm (1in).

Above: *Atractomorpha* grasshoppers are called '**coneheads**'. Here members of an Australian species are mating. 2.5–3.5cm (1–1.4in).

Deserts may seem unlikely places for grasshopper speciation, but after rare rain events, deserts bloom and long-storage grasshopper eggs hatch. Locusts are the most famous of these, but it's the invisible ones, the camouflage masters, which make desert species most fascinating. Here is a selection of cryptic species from Australia and Africa.

Top left: *Coryphistes ruricola*, a 5cm (2in) long grasshopper that copies bark well. Top right: Mating **rock grasshoppers**, *Raniliela* sp., 2.5cm (1in) and 2cm (0.8in). Above: A **sand grasshopper** that is invisible as long as it does not move.

The **grasshopper** here, in the family **Eumastacidae**, is flat and lying on its side among similar dead leaves. It is large in this picture, but finding it is still hard. Borneo, 3cm (1.4in).

Similar to the previous picture, *Chorotypus* sp., family **Eumastacidae**, is lying flat among rainforest leaf litter. Right: Note how thin and even dead-leaf crinkled, its body is. Only the head has depth, to accommodate the important feeding and sensory organs. From Sumatra, 2.5cm (1in).

Above: A rare master, a **grasshopper** that adapts to the hues of the rocks around it, here copying red marble perfectly, 3cm (1.1in). Bottom: A nymph of *Goniaea* sp., not only looking like a dead leaf, but behaving like one, by laying flat on the ground, 2cm (0.8in).

The dusty red soil of South Africa's west-coast deserts is home to a **grasshopper** in the family **Pamphagidae**, which even matches the dusty texture. 1.5cm (0.6in).

Two weird species to end the chapter. Left: The **pinhead grasshoppers** are a group in the family **Tetrigidae**, found in the forests of New Guinea. 2.5cm (1in). Right: *Bullacris* sp., the **bladder grasshopper**, family **Pneumoridae**, has this huge air chamber used to amplify its chirps, which can be heard up to 2km (1.2 miles) away. From South Africa, 4cm (1.6in).

STICK INSECTS
and LEAF INSECTS

Order PHASMIDA
(PHASMATODEA)

3,000 species in 3 families

This insect order is famous for most of its species living a hidden life. This, despite most of them being large to very large. They are the masters of crypsis, mimicking sticks, bark and leaves, sometimes with perfect imitation, down to leaf veins and damaged leaves.

The majority are the stick insects, in two families. Long, skinny, rounded, stick-like bodies vary in how closely they resemble actual sticks. However the most important part of hiding is not just looking like something, but behaving like that something. To achieve this part of the deception most stick insects are only active under the cover of night. During the day, they either keep perfectly still, or gently sway their bodies to imitate wind rustling the vegetation. If spotted by a predator, they either stiffen the body and fall to the ground like a real stick, or, if winged, suddenly unfurl bright wings in a startling display.

One problem of good camouflage is that when they are spread apart, the males have a big job finding the females. Many species have wingless females, but males are usually winged and do the finding. If a female does not meet a male, some species lay eggs without mating, which produce perfect copies of the female — a process known as parthenogenesis. The eggs,

100 to 1,000 or so, are also cryptic. They look like seeds, and are just thrown to the ground, not far from the food plants needed for the young. This extra camouflage works well in general, but ants, which often forage for seeds, pick them up and store them underground or eat them. Eggs can lay dormant for one to three years, although in the tropics they tend to hatch quickly. Very skinny and long first stage nymphs uncurl themselves from inside and start to eat the plants above them. All phasmids are vegetarians. After five or six moults they are adult and the cycle starts again.

The true leaf insects, in the family **Phylliidae**, are confined to tropical Asia through to New Guinea. Their camouflage is startlingly real. Even away from a backdrop of other green or brown leaves, they still look like a real leaf. During the day, many even gently sway like a leaf in the breeze.

And back to size. The official longest insect in the world, *Phobaeticus chani*, a stick insect from Borneo, measures 56.7cm (23in) long with outstretched front legs (a natural position for stick insects). What is more surprising is that a contender for the crown was only discovered in 2006, in Australia. *Ctenomorpha gargantua* has been officially measured at 50cm (20in), but an unofficial record of 61cm (24in), exists. Right now, eggs of this species are being reared to see if a specimen-certified individual can beat the Bornean record. Either way, these measurements are amazing. A typical stick insect in the tropics is around 10–20cm (4–8in) long, and leaf insects about the same.

A typical **stick insect**. Not perfectly camouflaged, but from a distance still generally blends in, especially as it does not move all day. *Podacanthus* sp., Australia.

Behaviour is as important as appearance if you want to hide from predators. This **stick insect** in Borneo was disturbed. It instantly became a stick, dropped from the leaves it was eating, body kept rigid, and landed like a real stick would. It will remain in this pose for a long time until the danger passes.

About half the species of **stick insects** are the true stick imitators. This one in Borneo is disguising its body against a vine, holding out its middle legs like small branches.

Opposite: This blue-legged **stick insect** from Madagascar is day active, hiding under large leaves. 7cm (2.8in) long.

Having more than one defence strategy is a trademark of many **stick insects**. If the general crypsis of this species, sitting flat on the lichen bark, fails, it has another trick. Suddenly it drops to the messy ground and lands upside-down with its straight-line body bent twice at 90 degrees. The underside has different shades, matching the dead leaves here, and the new shape makes most hunters lose its outline and give up.

Blue is rare among most insect groups. In Australia, *Megacrania* species stand out against the *Pandanus* leaves they sit on. However, if disturbed they show that the blue was a warning as they squirt the attacker with a sticky milky fluid. To us it smells like peppermint, but presumably this is not a pleasant smell to potential predators. 8cm (3.2in).

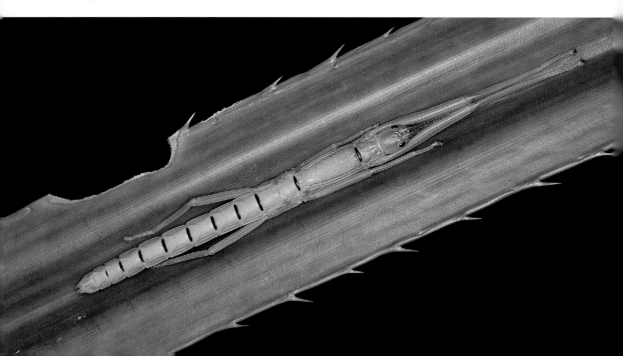

Not all **stick insects** are winged, and some are brachypterous, or short-sized. This stick insect in Madagascar cannot fly with these tiny wings, but suddenly unfurling them when disturbed, it may startle a predator long enough to make an escape.

Most stick insect males are different and very much smaller than the females. Australian species, female 140mm (5.6in), male only 95mm (3.8in) long.

Stick insect eggs are often mimics of seeds, and always little works of art. Here are four different species from Australia. The little 'lids' are just that. They pop open as the nymph starts to uncurl from inside.

This first-stage **stick insect nymph** in Costa Rica is all skinny body and legs. The egg it hatched from (see left) is quite squat, so a very convoluted origami folding process kept it inside until the day it uncurled itself through the egg lid. This species hides among white frass-like lichens which hang in the canopy.

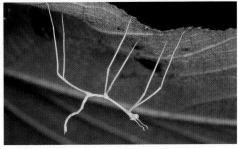

STICK INSECTS AND LEAF INSECTS

The final stick insect in this chapter, from Borneo, is near-perfect at concealment. There is one above, about 4cm (1.6in) long, mimicking not just the green but also the texture and pattern of the vegetation it is hiding on.

Opposite: The true **leaf insects** are large – this one from Indonesia is more than 10cm (4in) long – and yet are almost never seen. Their leaf resemblance is perfect, and combined with the habit of keeping still during the day, often high in the rainforest canopy, makes them invisible. This one genus, *Phyllium*, is found mainly in South-East Asia, and across to Madagascar.

Pyllium sicipholium literally means 'dry leaf'. A variation on the mainly green leaf insects, seen here away from a perfect backdrop for camouflage. 9cm (3.8in) long

WEB-SPINNERS

Order EMBIOPTERA

250 species in 8 families

These insects are unlikely to be seen by many observers. They spend their lives hidden in silken tunnels under bark and logs and other crevices. The small number of known species reflects this retiring life, and it is likely that there are many more to be found.

Web-spinners are mostly dark, elongate, cylindrical insects between about 0.5–1.2cm (0.2–0.5in) long. Males of many species have four thin wings, but females are always wingless. They spend their lives in the tunnels, created by silk, which unlike in spiders and silk-producing insects, is produced by their front 'feet'. Their tarsi, the last segments of their legs, have a bulbous silk gland in the first of three segments.

Another adaptation to the tunnel life is being able to run equally fast forwards as backwards. The wings of the males are also soft, and can be bent forwards over the head when moving backwards. When they need to be used for flight, dispersing to find females in other colonies, the wings are first hardened. After mating the females sometimes eat the males, as in some mantids. The young are nymphs, looking like small adults.

Above: A winged male **web-spinner** of the family **Oligotomidae**. When the males perform dispersal flights, they sometimes come to lights, as this Australian species did. Males are short lived, and sometimes do not feed at all once they are adults.
Left: A typical female wingless **web-spinner**, *Australembia* sp., family **Australembiidae**, from Australia.

ZORAPTERANS

Order ZORAPTERA

30 species in 1 family

This is a tiny order with tiny species unlikely to be spotted by casual observers. They are mostly less than 0.3cm (0.1in) long, and live in rotting logs and other out-of-sight moist habitats. Their claim to fame is perhaps the fact that we still cannot place them properly into the evolutionary lines of near neighbours. Features of bugs, web-spinners, cockroaches, lice and psocids are all mixed in these simple forms. Two general types of species are found, despite both being in the one world genus. All have moniliform antennae, meaning 'like a string of beads', but about half the species have no eyes or wings. And the other half do have small compound eyes and two pairs of short wings. Living in rotting organic habitats is perfect for their diet of mainly fungi, with the odd mite apparently also taken. Their modern distribution encompasses the Americas, Africa and southern Asia – there are none in Australia or Europe.

Two minute **zorapterans,** both about 0.3cm (0.1in) long, in the family **Zorotypidae.** Above: *Zorotypus gurneyi* from Costa Rica. Opposite: The almost see-through *Zorotypus hubbardi* from North America.

BOOKLICE

Order PSOCOPTERA

3,500 species in 26 families

The booklice get their name from a tiny common species that shares our houses and stores. Its habit of eating certain glues and paper products make it a pest of books. In general booklice are soft-bodied small insects, from about 0.5–10mm long (0.04–0.4in). They have biting mouthparts and their food consists of myriad different plant, fungal and detritus matter. None are hunters, although they are a favourite food of many predators, especially the tiny pseudoscorpions. Unlike the very bulbous wingless true 'book' lice, most species in the family are more elongate, winged, and live out in natural habitats such as forests. They can be found under leaves, under and on bark, in leaf litter and in other hiding places. A few species can spin silk, such as the Embiopterans, and they use this to create shelters on bark. In the tropics, the drab species are replaced by some very showy ones, with bright long-winged bodies. Several families have communal species, living in groups of up to hundreds, often in the open, displayed on tree trunks.

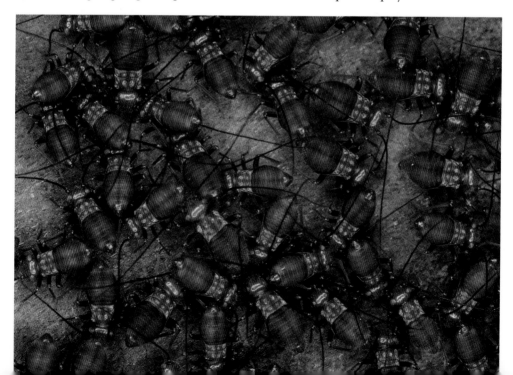

The life cycle consists of eggs laid in hiding, followed by five or six nymph stages before the often winged adult form. This group is closely related to the true lice (next chapter), which have species with both biting mouthparts, and the more famous sucking mouthparts. Even though some booklice can be found in or near bird and animal nests, they are only there for the detritus, and no species are parasitic like the true lice.

Above: A solitary large **booklouse**, *Sigmatoneura formosa*, family **Psocidae**, from Australia. Winged forms hold their wings roof-like over the body, quite pointy at the end, and usually have long thin antennae.

Opposite: The word lice does a disservice to this order. They are not parasites, and many are beautiful little insects, going about their business in the nutrient-recycling story in natural habitats. This communal species of **booklice** in Thailand is one such example. Each about 0.4cm (0.15in).

This Ecuadorian species is among the largest **booklice** in the world at 1cm (0.4in) long. With their red-and-black warning patterns, they live openly on top of leaves rather than hiding underneath. However they are just mimicking other insects which are poisonous, as booklice tend not to store poisons of their own.

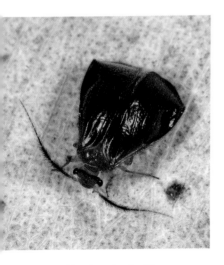

This Australian **booklouse**, *Calopsocus* sp., in the family **Calopsocidae**, is an exception to the rule. Its wings end in a flat line, rather than the usual point. It's a fungus feeder in eucalypt forests.

The cosmopolitan genus which gives the family its common name of **booklice** is *Liposcelis*, family **Liposcelidae**, and these can indeed be found in books, feeding on the glues and old paper.

The winged **booklice** are often very elongate with clear wings longer than the body. This is another communal species, living under rainforest leaves in Indonesia.

PARASITIC LICE

previously in the Order PHTHIRAPTERA, now merged with the Order PSOCOPTERA

3,500 species in 27 families

These are the true lice. Most are ectoparasites, meaning that they live on the outside of the body of their hosts, which are mainly birds and mammals. They evolved from a group which split into the booklice (previous chapter) and two major forms of parasitic lice. One group has chewing mouthparts, and mainly lived off the detritus of skin and feather bits in animal nests. Some booklice still have this habit. However, over time the true lice adapted to leave the dangerous habitat of the nests and live directly on their hosts. Here they lost wings and became more flattened and short legged. Some also rasped at the host's skin and fed directly on blood and oils. At some point another group evolved more stylet-like mouthparts, and these sucking lice are more likely to drink blood as the main meal.

Because of being adapted to the specialised environment of the skin of their hosts, moving between hosts is difficult. There is no developmental stage adapted for moving freely. Therefore, in general they do not leave the animal except when two hosts are in very close body contact. This means that most lice species are tied to only one, and rarely two host species. With about 3,500 species now known, it means that almost that many different species of birds and mammals have their own louse ectoparasite. Humans have three species: the hair louse, the body louse, and the pubic louse.

For some reason there seems to be a lack of photos of the human **Pubic Louse**, *Pthirus pubis*, family **Pthiridae**, in its natural habitat! The claws are among the largest of all species, adapted to hang onto hair with much tenacity. Luckily they are now much more rare than the other human louse species.

The short legs of lice have often developed large strong claws for hanging onto hair as the host is scratching to dislodge them. Bird lice have either claws or wrap their legs around bits of feather. These species are also usually more elongate and skinny, to better flit between feathers as the bird is searching for them with its beak. Among the most specialised species is the seal louse, which spends its whole life in the blubbery hair spaces near the seal's skin — where air is often trapped on long dives.

Eggs are laid on the most inaccessible body parts. In birds it is often the head and neck, where the bird cannot reach with its beak. Lice have a very strong 'cement' they exude with the eggs, gluing them to hair and feathers. They have to be careful during this process, as it is so quick drying, they can become permanently glued to the hair themselves. The life cycle is very quick, as little as eight days from egg to adult for the human body louse.

Typical feather-adapted lice. They are very flattened, with hardened slippery body plates, able to move between layers of feathers very quickly. These ones live on buzzards.

To put the habitat of lice in perspective, here is a **louse** species which lives on brush-turkeys in Australia. Note how very tiny it is on the scale of a detail of one feather. It makes them both relatively harmless to the host, and very hard to find and dislodge.

A bird louse species from North America, the **Slender Duck Lice**, *Anaticola crassicornis*, on a feather. More bird lice have this elongate shape than the rounded shape, although they are always very flat.

Above: Anyone who has watched any of the ape and monkey species, notices how very much time they spend grooming each other. It serves to bond the troop together, but in the process those deft little fingers are dislodging and eating the **lice**. Here Barbary Macaque in Morocco are deep in the process.

Left: This is the best known, and most hated louse. The human **Head Louse**, *Pediculus humanus*, family **Pediculidae**, is very hard to get rid of once a child brings them home. Unlike most lice they are very adept at moving between hosts, causing 'outbreaks' at schools and other meeting places. The hair shows their small scale and acrobatic skills.

TRUE BUGS, HOPPERS, CICADAS, APHIDS and SCALE INSECTS

Order HEMIPTERA

88,000 species in 131 families

What's in a name? In a book about world insects, for a world audience the word 'bug' needs to be set in context. In North America especially, 'bug' simply means all insects. This use is spreading, but the original meaning refers to insects in the order **Hemiptera**. A 'true' bug is defined by having a rostrum, a beak-like melding of all the mouthparts that most insects have separate. It has no jaws for chewing, and therefore is a sucking insect.

The 'true' bugs are usually winged, with the top wings leathery and forming wing-covers which cross over on its back. Beetle wing-covers are straight along the back, and they have chewing mouthparts. Rear wings are for flying and usually transparent. They grow as nymphs from eggs, undergoing five or more moults before becoming adults. The nymphs look like small adults, but can sometimes have different patterns, and a more squat shape in the early instars (stages). As their diet is the same, the herbivorous species are often seen in many stages on the same plant. The herbivores include some pests, as they suck plant juices for food. When the plant in question is a crop there is trouble, but these injurious species make up a tiny percentage of bugs as a whole.

This is a large and varied order, and so not all are herbivores. There are some fabulous hunters, especially the well-named assassin bugs. Without jaws, they pierce the prey with the 'beak', inject digestive fluids, and drink in the contents. Assassins are terrestrial, but most of the carnivorous families are water based. Eight families are fully aquatic, breathing by coming to the surface and storing air. Three other families are adapted to live on the surface tension on top of the water. These are the pond skaters, or water striders.

The use of the term 'true bugs' matters, as this order has three quite different divisions.

The other two are the hoppers and scale insects.

The hoppers include treehoppers, planthoppers, fulgorids, cicadas and many others. They are also often winged, and are usually elongate, with larger rear legs, which in some can be used to hop and jump. In fact, the most adept jumping insect in the world is not the flea, but a species of planthopper. Most of these are plant feeders, some living in family groups, others widely dispersed. The cicadas make extremely loud mating calls within their hollow abdomens, while most hoppers have mating calls too far towards the 'subwoofer' end of the scale for us to hear. Among them are also a few pests, as the name planthopper may warn.

The strangest insects of all are the scale insects and their kin. Their bodies are soft little blobs of usually clear flesh, with minute simple legs and a rostrum for drinking through.

None of this is visible to most observers, as the name 'scale' insects comes from the fact that they hide these delicate bodies under a variety of waxy, shell-like shelters — the scales. The females of most species never see the light of day once a shelter is established. The wax scale grows with the occupant as it drinks the plant juices non-stop. The males of many species do have a winged form to disperse and find females to mate with. This is the group with the most pests. Once established, they multiply very quickly and can slow a plant's growth markedly. The excess sugars they exude can attract sooty moulds (molds) and other fungi that further injure plants. And some have an unfair advantage in the world of tooth and claw. Ants love to drink the excess sugars that scale insects produce, and so they protect and farm these sweet food givers.

Other insects in this group include the aphids, mealybugs and whiteflies.

The True Bugs —
Suborder HETEROPTERA

The most classic bugs are the **shield bugs**, which are sometimes called **stink bugs**. Typical body form in this family, **Pentatomidae**, is oval with a bit of a point at the back. Most have flying wings under crossed-over wing-covers. *Nezara viridula* is a common pest of vegetable gardens almost all over the world. 1cm (0.4in).

Most **shield bug** species do not visit our gardens. Some are quite spectacularly patterned, and many produce foul smelling fluids when attacked. *Catacanthus nigripes* is a showy species from Australia and New Guinea.

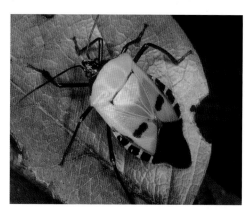

Catacanthus incarnatus is a warning-patterned **shield bug** species from New Guinea. This same yellow-and-black pattern is mimicked by some other insects and even spiders, all trying to benefit from the protection this poisonous bug enjoys.

Another family of shield bugs, the **Scutelleridae**, are mostly brightly patterned and differ by having wing cases that look like a full body shield. *Tectocoris diophthalmus*, the **Harlequin Bug**, is among the few pests in this group. Here are new nymphs recently hatched from the eggs above, and ready to bother some ornamental plants and cotton.

The nymphs of the bug family **Tessaratomidae** are often more outrageously bright than the adults. Here are two species, *Lyramorpha rosea* from Australia (above) and another *Lyramorpha* species from Borneo (previous spread).

The **lace bugs**, family **Tingidae**, are flattened with see-through flanges larger than the body underneath, and lacey honey-comb patterns. This species from the genus *Corythucha* is from North America. 0.5cm (0.2in).

The striped **stink bug**, or 'minstrel bug', *Graphosoma lineatum*, is a common European species frequenting the celery and carrot family of plants. 1.2cm (0.5in).

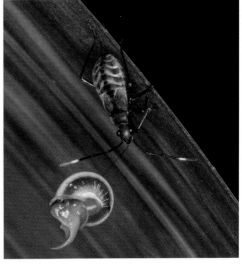

The Eurasian **shield bug** species *Euryderma ornata* has several morphs and specialises on sucking the sap from the crucifer plant family, Brassicaceae. 0.8cm (0.3in).

Bugs in the family **Lygaeidae** are often called **seed bugs**, but some are hunters. Note the amazing strength of this extremely thin rostrum it managed to pierce and hold the snail shell with.

Eight families of true bugs live fully aquatic lives. They are mostly hunters, living under or on top of the water. They have not evolved gills, so need to come to the surface to replenish air, or breathe through a modified syphon, like a snorkle. Most can fly away to find new ponds when their habitat dries up.

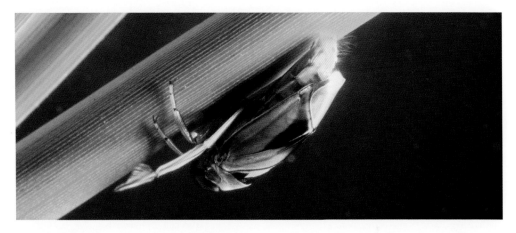

The **backswimmers** are so called as they swim 'on their back', or upside down. The rear legs are oars that propel them in very fast spurts. *Enithares* sp. from Australia, **Notonectidae.**

The largest water bugs are the **giant fishkiller bugs**, family **Belostomatidae**. This Australian species is more than 7cm (2.7in) long. It uses its 'ice-tong' front legs to impale passing prey such as tadpoles and fish, and sucks out their contents through tubular mouthparts.

The **Belostomatidae** family also includes smaller **water bugs**, such as the one shown here, which is sucking out the contents of a water snail. This group is famous for the females gluing their eggs onto the back of the males, so they can protect them before hatching.

These sleek water acrobats in the family **Gerridae** have many names, such as **pond skaters** and **water striders**. Their very long legs have special mats of fine hairs that resist water and enable them to live on the surface. The middle legs are used to 'oar' their way along at very high speed. This one, *Tenagogerris euphrosyne* from Australia, is eating a dead cicada on the surface.

This large **pond skater** lives on rainforest ponds in South-East Asia and New Guinea. Its legs pick up ripples made by insects on the water, creating a picture of its habitat like returning radar waves. Family **Gerridae**, body 2cm (0.8in) long.

This bright **assassin bug**, family **Reduviidae**, from Ecuador is a mimic of local bees. Not all predators like to tackle social insects such as bees and ants as they can be good at retaliation. Therefore this outer deception serves somewhat to protect this insect in the 'eat or be eaten' world it inhabits.

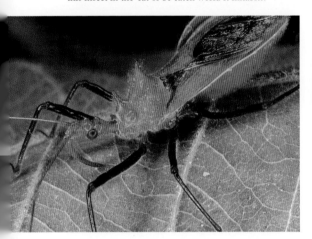

Assassin bugs, family **Reduviidae**, are the most common hunting bugs in most habitats. The curved 'beak', or rostrum, stabs the prey and injects digesting fluids (right). It then sucks out the contents. Most are quite showy and are found in the open, but a few species use camouflage for extra stealth. The brown mess (above right) is an assassin nymph from Ghana, covered in the mud of the termite nest on which it hunts termites.

Members of the family **Coreidae** are sometimes called **squash bugs**. Most are robust and brownish, but *Anisoscelis affinis* in Costa Rica stands out from the norm.

The bug family with the most species is the **Miridae**. Most are elongate, with long antennae and virtually any pattern combination. Some are crop pests, but most live their lives away from us, like this Malaysian species pollinating a flower.

The family **Pyrrhocoridae** contains orange and red species that are often seed-eaters. They liquefy and suck out the contents, sometimes in a frenzied group like these *Dysdercus* sp. bugs from Madagascar.

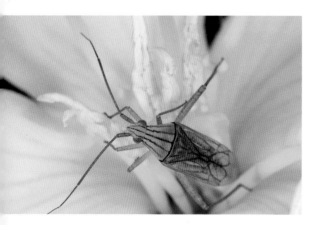

The Hoppers, Cicadas — and the Scale insects, Aphids and others, Suborder Homoptera

Cicadas generally spend two or more years underground feeding on roots. Their tough digging-adapted nymphs then climb up a tree in the night, and the winged adult form emerges. Here an Australian species has just emerged, and is still drying its body and wings before it can fly away.

Below: The most famous cicada is the so-called **17-year Periodical Cicada** (sometimes misnamed as the 17-year locust), found in North America. The nymphs, which dig tunnels around tree roots, live for 17 years before emerging in huge numbers. The noise of the calls to mate can be loud enough to be painful to the human ear, as the short-lived adults mate and lay eggs for the next generation.

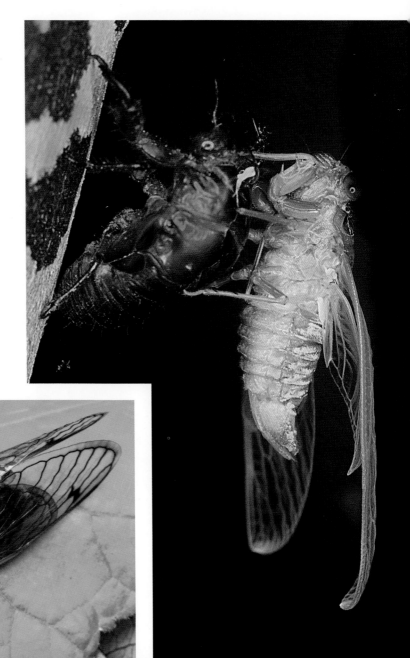

Cicadas make their sound by vibrating a drum-like skin inside a hollow chamber in their abdomens. A few species are called '**double drummers**' as they have two drums. This otherwise undistinctive species, *Thopha sessiliba* from Australia, is one of the loudest insects in the world.

In order to not attract a predator while making a very loud noise to attract a mate, some **cicadas** use a camouflage setting. This is *Yanga auttulata* from Madagascar. 5cm (2in) long.

One of the largest **cicadas** in the world is also one of the most striking. *Tacua speciosa* lives in the rainforests of South-East Asia. Its wingspan of up to 18cm (7in) is very impressive. It is actually a little bit larger than the scale of the picture on this page.

The highland forests of Thailand are home to many eye-catching **cicadas**, such as *Heuchys fusca*. 5cm (2in) long.

The largest group within the Homoptera are the planthoppers. They are elongate, often flattened insects with large rear legs which in one species hold the record for the longest hop of any insect. They are infamous for a few species being pests of crops, but the majority of species are harmless and often spectacularly beautiful insects, mainly in the family **Cicadellidae**, which alone has more than 20,000 named species.

Macunolla sp., from Costa Rica, **Cicadellidae.**

Tiny 0.4cm (0.16in) **hopper** from New Guinea, **Cicadellidae**.

Hopper from Costa Rica, **Cercopidae**. 1.4cm (0.6in).

Hopper from Madagascar, **Cicadellidae**. 0.8cm (0.3in).

Hopper from Costa Rica, **Cicadellidae**. 1.5cm (0.6in).

Agrosoma sp. nymph, Costa Rica, **Cicadellidae**.

Ladoffa dependens, **hopper** from Costa Rica, **Cicadellidae**. 0.8cm (0.3in).

Hopper from Madagascar, **Ricaniidae**. 1.4cm (0.6in) across.

A flatid **hopper** from Indonesia, **Flatidae**.

Hopper from New Guinea Highlands, **Cicadellidae**. 1.2cm (0.5in).

Bhandara sp. from Malaysia, **Cicadellidae.**

Agrosoma sp. from Costa Rica, **Cicadellidae**. 1cm (0.4in).

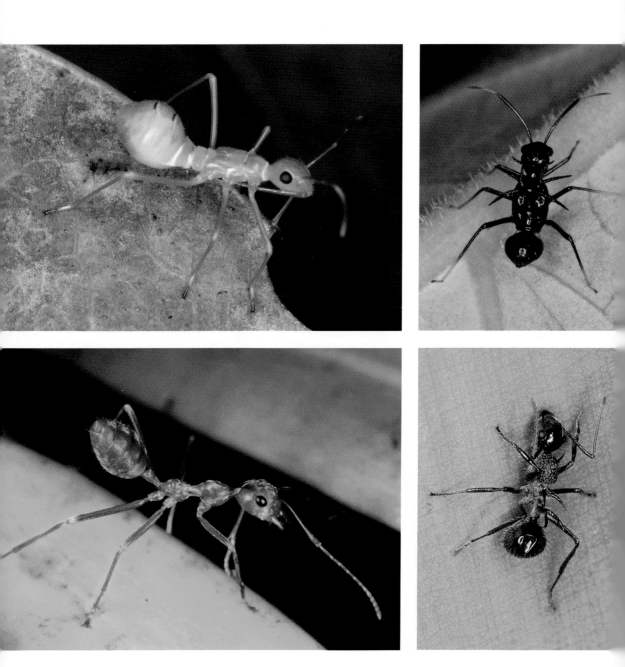

A few of the bug families with elongate bodies mimic ants, either in the first instar or all preadult stages. Ants can be tough prey as they defend their own in numbers, so looking like them can help another insect to live longer. Left: The green bug (top left) is in the family **Alydidae**, and mimics the notoriously angry green tree ants, *Oecophylla* sp. (above left) from Africa to Australia. Right: The black bug (top right), also an **Alydid**, mimics any of a huge tree ant genus, *Polyrhachis*, most of which have the detail of the spines on the back (above right).

Planthopper of the genus *Macugonalia*, family **Cicadellidae**, from Ecuador. 1cm (0.4in) long.

The **Rhododendron Leafhopper**, *Graphocephala fennahi*, is a native of America where it sucks the sap of rhododendrons. It has been introduced to Europe where it is now very common.

Planthopper in the **Cicadellidae** family from Malaysia. 1.2cm (0.5in) long.

Hopper nymphs can be far more strange than the adults. Many use waxes from plants to create elaborate crazy disguises that are constantly growing and changing. Shown here are two flatid nymph species, from Madagascar on the left and Borneo on the right.

The family **Issidae** contains tiny domed hoppers that seem to mimic beetles. This is a species in the genus *Hemisphaerius* from Indonesia.

The **Derbidae** family has some very odd long-winged species in the tropics. This one from Madagascar. Wingspan 2.5cm (1in).

Borneo is the key location for **fulgorid bugs**, with many large and bright species sitting on rainforest tree trunks. Family **Fulgoridae.**

Another Bornean **fulgorid bug**, *Fulgora* sp., 4cm (1.6in).

The **lantern bugs** are weird enough without their name coming from a misconception that the bulbous 'nose' was lantern-shaped and glowed. Their more apt common name is the peanut-headed bug. This large 6cm (2.3in) hopper is the original species named in South America – *Fulgora laternaria*, in the family **Fulgoridae**. It can grow up to 9cm (3.5in).

This very pale **hopper** with wax frass protruding out of its rear is in the family **Cixiidae**. It lives a blind dark life in lava tubes in Australia.

The common fulgorid in Borneo is this handsome 5cm (2in) beast in the genus *Fulgora*, family **Fulgoridae**. It is even more noticeable with the bright wings outstretched. Most hopper families are winged, although many species prefer walking and jumping to flight.

The elaborate 'head gear' of many **membracids** can be larger than the rest of the body – which is just the little bit behind the wings. The bug would probably have a hard time balancing if not for the fact that these showy structures are largely hollow. Shown here is a species in the genus *Heteronotus* from Ecuador. 1.2cm (0.5in).

The bug family **Membracidae**, the treehoppers, are famous for elaborate shapes that have no know function. They often live in family groups with their bright and differently shaped, nymphs.

A family group of **treehoppers** from the genus *Antianthe* in Belize. Note the tiny red, spiny and seemingly unrelated nymph at the bottom.

Cladonota luctuosa, a species from Costa Rica, where the diversity of the **Membracidae** hopper family is at its highest.

The **treehopper** *Poppea diversifolia* from Costa Rica does not stop at odd shape, but adds wild markings too. 1cm (0.4in).

Leptocentrus taurus from Malaysia. An amazing display of head gear on a tiny 0.8cm (0.3in) long member of the family **Membracidae**.

Central American **treehoppers** in the genus *Umbromia* mimic spines on acacia plant stems.

Stictocephala bisonia, a very triangular **treehopper** species from Costa Rica. 1cm (0.4in).

An unidentified **treehopper** from New Guinea, 1cm (0.4in) long.

This final group of bug families used to have its own suborder, as it starts with the recognisable aphids, but then includes the very weird scale insects and their kin, which mainly consist of species of soft blobs hidden under elaborate wax shell or scale-like structures.

Only a fraction of aphid species worldwide are pests in our gardens, but this one is probably the best known. **Rose Aphid**, *Macrosiphum rosae*, in the family **Aphididae**.

Aphis nerii is an **aphid** with an almost worldwide distribution. It has travelled with garden plants such as sweet potato, sorghum and soya as humans have moved them around.

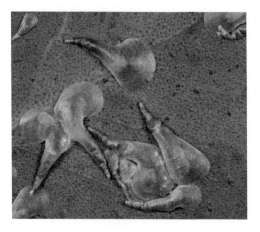

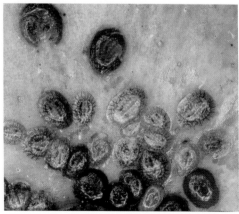

Members of the family **Psyllidae** are known as **lerp insects** – 'lerp' being the term used for the wax shells they live under. In this image of a member of the genus *Creiis*, on eucalypus leaves in Australia, the tiny 0.2cm (0.08in) creature is just visible inside as the orange blob.

This is one of the infamous **scale insect** pests, the 'soft brown scale', *Coccus hesperidum*, family **Coccidae**, here attacking a mango.

THRIPS

Order THYSANOPTERA

6,000 species in 9 families

Thrips are minute to small insects, from about 0.5–4mm (0.02–0.16in) in length. Some species are pests of crops and ornamentals, although as always in the insect world the majority of species do us no harm. Most eat flowers and plants that don't affect our livelihood. Many feed on fungal spores and many are predators, sometimes helping us by hunting insects such as aphids, which are pests.

Their body design is like no other. As a general rule, all multi-celled organisms are symmetrical. However thrips have lost one mandible, and the other mouthparts are adapted into a short tube. To feed on plants they pierce a cell with the single mandible, then insert the tube and suck out the contents. Bugs, which also suck on plants, insert their longer, stronger tube deep into the plant and suck in the phloem or xylem – the juices flowing up and down in the 'veins' of plants. The difference is that thrips kill the cell they invade, while bugs slightly weaken the plant as a whole by depriving it of some of its sugary fluids. The few genuine pests are also dangerous as they can carry plant diseases between plants, which then travel the world with our crops.

Another way thrips are unique is their life cycle. They have two nymph stages, followed by two or even three stages that are essentially pupa. Thus they are somewhat between the incomplete and complete metamorphosis groups of insects. This cycle can be very short, as little as three weeks, with many generations in a year, especially in the tropics.

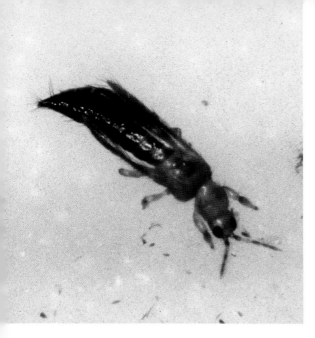

The Hawaiian **flower thrip**, *Thrips hawaiiensis*, is very typical. At 0.3cm (0.1in) it is tiny, but it has spread quickly around the world on a variety of plants. Here on bananas.

Thrips are at their worst when they coevolve with a virus to be its carrier. These are symptoms of the Tospovirus, attacking tomatoes via thrip bites.

The family **Phlaeothripidae** stands out by having the largest **thrip** species. At 1.2cm (0.5in) this *Mecynothrips* species from Australia is a monster among thrips. It feeds on fungus.

The majority of thrips look rather dull, but the tropics have some wildly bright species, such as these 0.3cm (0.1in) **fungus thrips** from Indonesia.

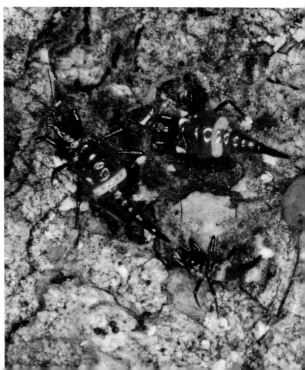

ALDERFLIES and DOBSONFLIES

Order MEGALOPTERA

300 species in 2 families

SNAKEFLIES

Order RAPHIDIOPTERA

260 species in 2 families.

The very small order Megaloptera is closely related to the even smaller order, the Raphidioptera, and to the large Neuroptera order (the lacewings and antlions). All three originate from similar ancestors as far back as the Permian period up to 200 million years ago. They are the earliest insects to possess a pupal stage, undergoing full metamorphosis from larva to pupa to adult.

Larvae of Megaloptera are aquatic, and often very long lived, up to five years in cold areas of the Northern Hemisphere. They are predators and have long tendril or feather-like gills along the back of the body. These can be pulsed in and out, to draw in more oxygen in waters with poor flow. Anglers know these creatures' larval stages well. The large and ferocious larvae are known as 'hellgrammites' and make very good lures.

When ready, the larvae leave the water and pupate in soil close by. Emerging adults are short lived and often do not feed. Many species are large, reaching a wingspan of 16cm (6.5in) in one American species. This is also a very odd species, as the males have huge tusk-like jaws, although these are only for ritual combat with other males. Lesser tusks are also a feature of the males of many other species.

The north of the Northern Hemisphere is the key location for both groups. The snakeflies have terrestrial larvae, which hunt other insects under bark. Females lay their eggs under the bark using an ovipositor, or egg-laying tube. Snakeflies are rare in the Southern Hemisphere, and absent from Australia.

The monster of the dobsonflies is *Corydalus cornutus*, the **Eastern Dobsonfly** of North America, with a length of 7.5cm (3in). The huge tusks are weak and used for ritual combat only.

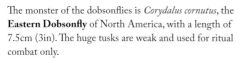

Alderfly larvae are aquatic hunters, equipped with these seven pairs of gills. Anglers know them as hellgrammites.

Although not tusked like many of the northern species, this Australian **dobsonfly** is large, with a wingspan of about 10cm (4in). Like all **megalopterans**, it is a slow and clumsy flier.

Snakeflies are most common in Europe and North America. Their heads are mounted on a long hinged neck, and together with a long slim ovipositor at the other end, this makes the name apt. *Phaeostigma notata* is from Europe.

LACEWINGS
and ANTLIONS

Order NEUROPTERA

5,000 species in 14 families

The Neuroptera comprise several well-known groups, with names such as antlions, owlflies and lacewings. The literal translation of the name means 'net-wings', and their large wings have complex venation that does look like a net.

They are related to the alderflies, Dobsonflies and snakeflies (see previous chapter) and share their history of being one of the first groups to evolve the complete metamorphosis method of growth – larva to pupa to adult.

Perhaps the best-known members are the antlions, the **Myrmeleontidae**, with larvae which live in sand pits, devouring ants which fall into the soft-walled traps. The lacewings, with a family of green species (**Chrysopidae**) and a family of brown species (**Hemrerobiidae**) are well known too. Many come to lights and the green lacewings as sometimes called 'golden-eye' for having startlingly gold patterned compound eyes.

The larvae of all families are predators. Most are free-living, roaming in search of insect prey. Many, such as the lacewings, attach the dead bodies of their prey to their own bodies, resulting in a walking mess of body bits that is unrecognisable as a living creature. Presumably this serves to confuse any potential predators of the larvae, which are otherwise soft and vulnerable. The larvae have fused mouthparts, resulting in tubes through which they suck the contents of their prey. The resulting empty husks are lighter to carry on their backs too. Two families have aquatic larvae. The small **Sisyridae** have gills and feed on freshwater sponges. The other family, the **Osmylidae**, have semiaquatic larvae with no gills, but which spend much time under water hunting mainly midge larvae.

Lacewings in general are more common in the tropics, and the antlion group is most speciated in dry habitats, where their sand-pit traps are easy to build and maintain.

Not all adult lacewings feed, but many are hunters on the wing, like the dragonflies. The predatory nature of these insects, and especially the preference of many lacewing larvae for insects such as aphids, makes this group a great friend to gardeners.

Last but not least is an example of convergent evolution. The praying mantids were not the first to evolve raptorial forelegs.

A marine shrimp beat them to it. But among the Neuroptera is another example of this hunting method, within the family **Mantispidae**. These 'mantis flies' have the long necks and very mobile heads, and of course the raptorial forelegs of mantids. Nevertheless their general appearance is still more lacewing than mantis.

The name **lacewing** is most applied to the family **Chrysopidae**, which contains mainly green, long-sized insects, often with beautiful eyes. For some, like this Australian species, the common name is 'golden eyes'.

Owlfly larvae are free ranging and have outrageous jaws. These are used to impale prey, deliver digestive fluids, and suck out the contents. 1cm (0.4in).

Green lacewing larvae, family **Chrysopidae**, are elongate with hypodermic jaws that are used to grasp, digest and suck. Their favourite prey is aphids, making them very welcome in gardens. The mess on its back are dead aphid bodies, which are stored for camouflage.

The **owlflies**, family **Osmylidae**, have big eyes, and this peculiar way of sitting with the rear end extended away from the wings. Their larvae have both terrestrial hunter species (above left) and aquatic ones, hunting midge larva.

The commonest lacewings are the **antlions**, family **Myrmeleontidae**, which are famous for the sand nests of the larvae. Adults are usually clear-winged but this African species is a beautiful exception to the rule.

The small family **Nymphidae** contains large showy species with larvae which hunt in leaf litter. *Nymphes myrmeleonides* is from Australia.

This beautiful East African **antlion** is one of more than 60 species in the genus *Palpares*, family **Myrmeleontidae**. Wingspan more than 10cm (4in).

These conical sand pits are known to most. However few people have seen the creatures within. The larvae of **antlions**, family **Myrmeleontidae**, flick sand up to create the unstable cones. When an ant falls in, the giant jaws of the antlion (right) await at the bottom. They need very dry sand, so often use rock overhangs or set up under houses.

Like all hunters – think mantids and dragonflies – the face of an **antlion** is equipped with large and wide-field eyes, and a very inquisitive stare.

These mating **owlflies** are so iconic in central European mountains that they have been given common names like butter-flies – this is the 'owly sulphur'. Their large bright bodies, with extended wings like dragonflies, can be seen chasing prey on the wing. The larvae are also hunters with their huge jaws. This species is *Libelloides coccajus*, body length 2.5cm (1in).

Opposite above: **Spoonwings** tend to live in dry areas where their larvae hunt in dusty or sandy places and overhangs. *Chasmoptera* is a genus of the more showy Australian species.

Opposite below: The **Mantispidae** family has independently evolved the raptorial forelegs shared by praying mantids. They are used the same way, to strike at prey and hold with barbed edges. **Mantis fly** eyes are often opalescent, as in this species from Australia.

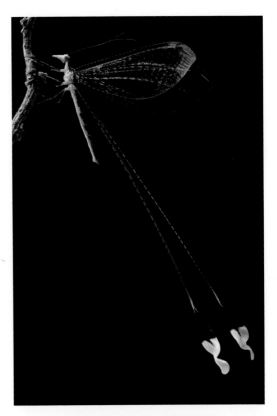

Africa is the key location for the evolution of lacewings, with many of the most extreme forms. The family **Nemopteridae**, the **spoonwings**, are especially beautiful. Shown here are *Nemophistha contumax* (above) and *Nemeura longstaffi* (below), both from Mozambique. The tailed rear-wings disperse pheromone and are each up to 6cm (2.4in) long.

A large **mantis fly**, family **Mantispidae**, with general wasp-mimicking characters from New Guinea. Note the toothed, powerful front hunting legs. *Euclimaciella nuchalis*, 2.4cm (1in) long

This **mantis fly**, family **Mantispidae**, from Uganda is 1.5cm (0.6in) long. There are more than 400 species of mantis flies in the world.

BEETLES

Order COLEOPTERA

420,000 species in 174 families

As the numbers above suggest, this is a very large group. Nearly one third of all described animal species on the planet are beetles. The basic body form, of a tough exoskeleton with equally tough upper wings covering the flying wings, has been extremely successful. With more than 170 families, there are no potential lifestyles that beetles have not adapted to. Herbivores, carnivores, scavengers and parasites can be found in most areas outside of Antarctica.

The beetle body is easily adapted to extremes, as it is very protective both of physical damage and of dehydration. The world's deserts teem with beetles in seemingly impossible places. The wing-covers shut in the rows of breathing tubes, the trachea, to minimise water loss.

Ground beetles, family **Carabidae,** are a large family of hunters. The majority are ground runners, chasing prey mainly at night. The scary-looking *Mecynognathus damelii* is the largest species in Australia, at 6cm (2.4in).

Conversely, many families have adapted to water, using the wing cases to create an air chamber which protects the trachea from contact with water. They only need to expose the tip of the abdomen to air sometimes to fill this chamber with fresh air.

The variation in size among so many species is astounding. At one end are microscopic species in the family **Ptiliidae**, which at 0.25mm (0.01in) are so light that they have feathers for wings. Beating solid wings would produce too much thrust for creatures so light, as they almost float anyway. At the other extreme, several species vie for the prize of biggest insect. The goliath beetles of Africa are the heaviest, with rectangular brick-shaped beetles weighing up to 100g (3.5oz). That is four times the weight of an average mouse. These stout beetles reach about 7.5cm (3in) in length, but the longest beetles are the Hercules beetles, with a horn as long as the rest of the body, which reaches 20cm (8in) in total!

Beetles undergo full metamorphosis, with a larva to pupa to adult sequence. This usually means that the larvae have a different lifestyle and diet to the adult, making better use of available resources. In some cases, both stages share the same life, especially among the leaf-eating families.

Beetles are so visible and recognisable that they have more individual common names than any other group. These include ladybirds, scarabs, click beetles, ground beetles, tiger beetles, elephant beetles, long-horned beetles, flour beetles, dung beetles and so many more. One family stands out from the rest. The weevils are the most speciated group among the most speciated order. More than 60,000 named species of weevils live a largely herbivorous life in most habitats. Something about the plan of a hidden larvae eating roots, wood and seeds, combined with the iconic long-nosed (rostrum), hard 'shelled' adult form, has led to the most successful animals of all. Here is a tiny introduction to the beetles, arranged roughly in evolutionary sequence starting with the ground beetles and ending with the weevils.

Above left: Members of the genus *Calosoma* are among the most recognisable of **ground beetles**, with nearly 200 species, mainly in the Northern Hemisphere. These night hunters are caterpillar specialists, like this 3cm (1.2in) *Calosoma scrutator* from the USA, which is known as the 'fiery searcher'.

Above right: A small African **ground beetle**, part of a chain of mimic species across wasps, beetles and bugs. *Graphipterus* sp., Mozambique, 1.2cm (0.5in) long.

Right: Not all **ground beetles** have fearsome large-jawed forms. This delicate-looking species from Australia hunts on foliage rather than on the ground, and is feared by the caterpillars it chases. 1.2cm (0.5in).

Opposite: A **ground beetle** at a kill is not a pretty sight. This *Ametroglossus* sp. is making a mess of an earthworm in Australia. 2.2cm (0.9in).

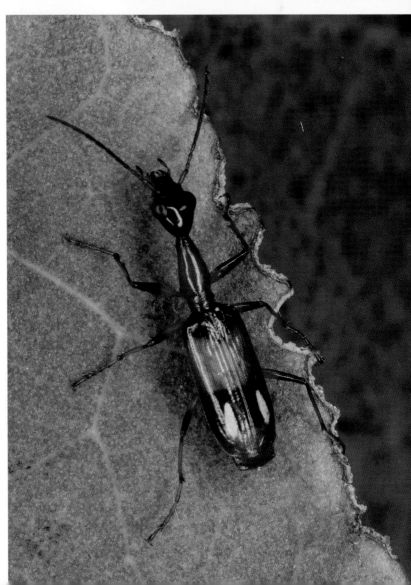

The largest genus in the **Carabidae** family is also the most beautiful. More than 900 species of *Carabus* are found in Eurasia, and many of these are very eye-catching.

While many are black, even these species may have a thin bright metallic margin. Some are shiny smooth, like *Carabus splendens* from Europe (opposite). The deep blue form (above) is a *Carabus* species from the Crimea region.

All species are night hunters, specialising in worms, snails and caterpillars. Species vary in size from 1.2–5cm (0.5–2in).

These beetles have found their way into human culture, as sightings are memorable. Many countries in Europe and Asia have used them on postage stamps. Shown here are examples from Hungary, Belarus, Poland and Russia.

Barbieri

Right: The inescapable jaws of the rainforest **tiger beetle**, *Therates fasciatus*, which hunts on leaves from New Guinea to India. 2.2cm (0.9in).

Below: The **tiger beetles** used to be placed in their own family, but are now treated as the subfamily **Cicindelinae**, which is incorporated into the **Carabidae** (previous pages). All are hunters, mostly very long-legged and fast-running, and with scimitar-like jaws. This is *Pseudoxycheila tarsalis* from Costa Rica.

Bottom: Most **tiger beetles** have wings and sometimes fly, but the genus *Tricondyla* has large wingless species running about tree trunks in tropical forests from Australia to India.

This extra long-legged and pale **tiger beetle** is adapted for life on hot beach sand, where it is so fast that it out-runs sitting flies before they can take off. *Cicindela* sp. from South-East Asia. 1.4cm (0.5in).

Below is *Mantichora latipennis*, from Mozambique, nearly cutting its large grasshopper prey in half with its extra long jaws, 4cm (1.6in).

This mating pair of **tiger beetles** is from the central savanna areas of Madagascar.

The rainforests of South-East Asia are home to this large-jawed **tiger beetle**, *Therates* sp., which hunts on leaves. Borneo, 1.8cm (0.7in).

Cicindela sexguttata is a common North American forest-dwelling **tiger beetle**.

Cicindela aurulenta is a common species in sandy heath country in South-East Asia. It is the largest winged in Borneo at 1.6cm (0.6in).

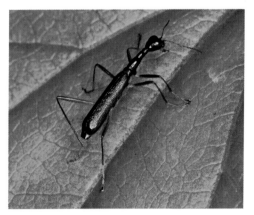

One of the less formidable species of **tiger beetles** – the skinny *Neocollyris* sp. from Thailand, 1.8cm (0.7in) long.

The largest **tiger beetles** in the world belong to the flightless hunters in the genus *Mantichora*. They possess huge, no-nonsense tough jaws and hunt mainly at night. Above is one scavenging a freshly run-over millipede in South Africa, 3.5cm (1.3in).

At the other end of the **tiger beetle** scale are these minute hunters in a moss forest in Borneo, at 0.6cm (0.25in).

A mating pair of ferocious southern African **ground beetles**, *Anthia circumscripta*. This genus contains many large night hunting species. 3cm (1.2in) long.

Bombardier beetles are nature's most exotic chemical factories. Two chemicals are held in separate sacs near its rear end, and when threatened the beetle allows these to mix, which causes a super-heated (about boiling point) corrosive explosion and smoke cloud. If this was not enough, it can also direct the turret-like release organ to spray an attacker from almost any angle, including forwards, under its raised body. This 1.5cm (0.6in) long species from the genus *Chlaenius* is from Mozambique, but other species are found in most parts of the world.

There are seven families of aquatic beetles, and the largest is the **Dytiscidae**, the **diving beetles**, with hunters as larva and adults. They store air under the wing cases for long dives, and use the hairy oar-like back legs for fast swimming. This member of the genus *Cybister* is a 3.5cm (1.4in) Australian species known as the 'toe biter' by those who put their feet into its ponds.

This odd **ground beetle**, with its flattened antennae, is part of a group which live with ants, the **Paussinae**. Adults and larvae produce an ant-mimicking smell which makes them welcome, however they repay this by eating the ant eggs and larvae. Apart from the trickster smell glands, they can also defend themselves with hot explosions similar to the bombardier beetles (previous page). Adults leave the ant nest to find a mate. *Cerapterus iaceratus*, from Mozambique, 1.5cm (0.6in) long.

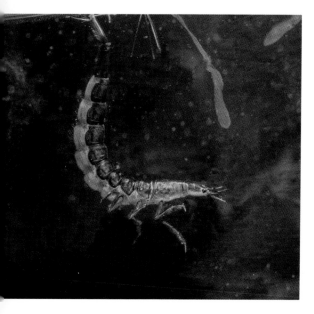

A **diving beetle larvae** comes to the surface to store air via the tail-like syphon. They are ferocious hunters of water insects and tadpoles. 2.5cm (1in) long.

Water beetles in the **Hydrophilidae** family are largely herbivores and scavengers. Their breathing strategy involves trapping air over fine hairs on the underside, and using the trapped oxygen slowly. This process is known as plastron respiration, characterised by the 'quicksilver-like' bubble. 1.6cm (0.6in).

Whirligig beetles, family **Gyrinidae**, gyrate very fast on the surface of the water. The waves this creates bounce off objects on the surface like radar, and shows them where potential prey and enemies are. Amazingly they also have four separate compound eyes – two adapted to focus through water looking down, and two above with focus modified for air. They are scavengers on insects that fall into water, and other drowned critters on the bottom. To dive, they carry an air bubble under and beyond the wing cases.

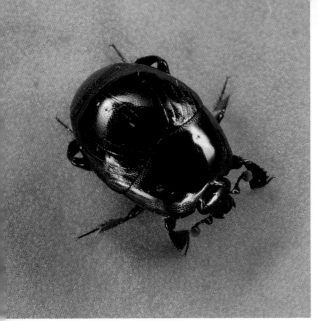

The family **Histeridae** contains mainly small oval beetles, living hidden lives near dung, carrion, in animal nests and even with ants and termites. Most are dark, but the genus *Saprinus* contains some beautiful metallic beetles.

Every beetle has its niche in life. **Burying beetles**, family **Silphidae**, bury dead animals by digging under them, and then lay their eggs in the carcass. Adults and larvae eat and recycle the flesh, like fly maggots.

The beetle family **Anobiidae** contains several stored product pests, many of which have now spread around the world. Most have 'normal' beetle shapes, but the **spider beetles** are quite weird. *Mezium affine*, an American species, is only 0.3cm (0.1in) long.

Not all members of the **burying beetle** family **Silphidae** bury dead animals. This European species, *Xylodrepa quadripunctata*, is a hunter of caterpillars. 2cm (0.8in).

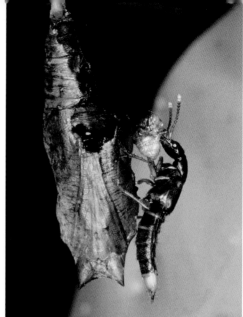

This skinny metallic **rove beetle** hunts soft insects, here searching for prey on leaves in New Guinea.

Rove beetles are a large family, the **Staphylinidae**. They are very skinny and elongate, with tiny wing cases that hide full-size wings. Most are hunters and scavengers, and this one is eating the parasite inside the pupa of a birdwing butterfly.

Although this North American **rove beetle** looks ferocious, it is actually a fungus-eater trather than a hunter. *Oxyporus rufipennis*, 1.2cm (0.5in).

Drama inside a roadkill rat. Flies are the first to arrive and their maggots eat the carcass, but soon maggot predators such as this **rove beetle** join the feeding frenzy. Beetle 0.8cm (0.3in) long.

The **Scarabaeidae** family is among the best-known around the world. It includes the chafers, 'June bugs', dung beetles, rhino beetles and more. More than 27,000 species are described around the world in 12 fairly distinct subfamilies. The following pages introduce examples from the most common to the most striking, starting with some of the largest beetles in the world.

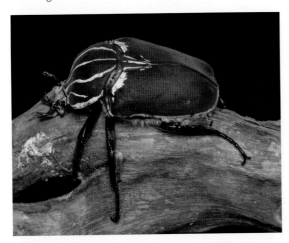

The **goliath beetle**, *Goliathus goliatus*, of equatorial Africa is officially the largest beetle in the world – not the longest, but the bulkiest. Its body can measure up to 11cm (4.3in) in length, but allowing for the large girth it can weigh up to 60g (2oz), while the larvae can be twice as heavy. They are in the chafer subfamily **Cetoniinae**.

This species of **goliath beetle** – *Goliathus cacicus* from Ivory Coast – is smaller than *Goliathus goliatus*, but when presented with a human scale it shows just how surprisingly large this genus of beetles are.

Another large **goliath beetle**, *Goliathus meleagris* from Zambia, 7.5cm (3in) long. Goliath beetles tend to keep to high forest flowers, such palm inflorescences, and rarely come to the ground. Their larvae live in rotting logs.

Opposite: One of the most splendid **goliath beetles**, *Mecynorhina polyphemus*, from equatorial African forests. 6.5cm (2.6in) long.

In Africa the **chafers** are known as fruit chafers, as apart from visiting flowers, they are attracted to sweet rotting fruit. Shown here is *Leucocelis haemorrhoidalis* from South Africa.

Chafers are often metallic, like this *Ischiopsopha* sp. from New Guinea. 2.8cm (1.1in).

In Australia *Eupoecilia australasiae* is known as the **fiddler beetle** for the pattern on its back. Large numbers of this **chafer** visit flowers in the spring.

Europe's **flower chafers** can be very pretty, like this *Trichius zonatus* with its furry red coat.

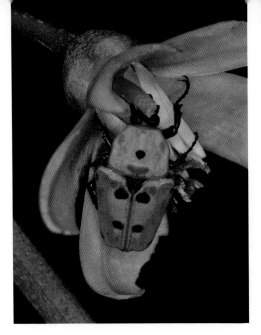

This **flower chafer** is from the eastern rainforests of Madagascar.

The furry furrows of *Trichaulax macleayi* from Australia, make this a very distinct **flower chafer**. 3cm (1.2in).

Polystigma punctata is a common **chafer** visitor to eucalypt flowers in Australia.

While the goliath beetles (previous pages) have the heaviest insect species in the world, the Dynastinae subfamily of the Scarabs have the longest. This **Hercules beetle**, *Dynastes hercules*, from Central and South America reaches 17cm (6.8in) long. The horns are used in ritual combat between the males, while the females are hornless.

An unidentified **flower chafer** from Mozambique displays the full spectrum of the rainbow. 1.5cm (0.6in) long.

Female **dynastid beetles** (hercules, rhinoceros) do not possess the horns that adorn most males. This species is from the New Guinea Highlands.

The three-horned male **Atlas Beetle**, *Chalcosoma atlas*, is one of the largest beetles in South-East Asia, at 12cm (4.8in) long.

The New Guinea highlands are also home to this three-horned male **rhinoceros beetle**, *Eupatorus* sp., 7cm (2.8in) long.

This is perhaps the best-known **Scarabaeid beetle**, the European *Melolontha melolontha*, the **Common Cockchafer**, which is on the wing in summer.

A tiny gold and silver speckled **cockchafer** from Malaysia. 0.8cm (0.3in) long.

The **melolonthid** group of scarabs, the **cockchafers**, are characterised by these stacked flat antennae, described as 'lamellate'. This species is from Thailand.

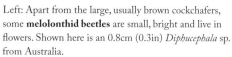

Above: Monkey beetle, Namibia, 10mm (0.4in).

Left: Apart from the large, usually brown cockchafers, some **melolonthid beetles** are small, bright and live in flowers. Shown here is an 0.8cm (0.3in) *Diphucephala* sp. from Australia.

In Africa, especially in the south of the continent, the huge profusion of wildflowers attracts many pollinators. There is a group of small **Scarabaeid beetles**, in the subfamily **Retulinae**, which often outdo even the bees — these are the thousands of species of **'monkey beetles'**. The males often have enlarged rear legs and claws, and at night many burrow into the flower for protection. Here some examples from South Africa. Most are only about 1cm (0.4in) long.

One of the literally thousands of **monkey beetles** pollinating the myriad flowers of the southern African flower hot-spots. Many are still without names. **Scarabaeidae**, 0.8cm (0.3in) long.

Asia has a small subfamily of **Scarab beetles**, the **Euchirinae**, in which the males have huge front legs, making movement ungainly for them. All species are rare. *Cheirotonus macleayi* is from Thailand, where it seeks fermenting palm-tree sap. Body alone 6cm (2.4in).

Gold is not common among insects, but gold beetles in the **Rutelinae** subfamily of **scarabs** have evolved on two continents. On the left is *Anoplognathus aureus* from Australia, and on the right is *Chrysina aurigans* from South America. The almost real gold sheen is produced by up to 70 layers of the chitin that forms their outer 'shell', each layer reflecting the gold individually.

The last **Scarabaeidae** subfamily, the **dung beetles**, are the quiet achievers of the ecosphere. They bury and recycle the droppings of most herbivore mammals and some birds, in the process fertilising the soil and lowering the numbers of pesky flies. They have been exported to places where species adapted to domestic animals like cattle did not exist, such as Australia. They are also the stuff of mythology, with the ball-rolling species seen as the mover of worlds in ancient Egypt.

The greatest of them all is the **elephant dung beetle**. The huge dung piles need a lot of work to distribute and bury. The **Addo flightless dung beetles**, *Circellium bacchus*, are up to 5cm (2in) long, and toil all day creating rough balls to push away and bury elsewhere. The male does the hard work, and often the female tags along but does not help. At the burial site they mate and she lays eggs that develop on this rich vegetarian fare. The beetles are very important and have right of way on national park roads, but unfortunately not all drivers respect this, and finding a flat massacre is common.

Mammal dung is not all **dung beetles** dispose of. This tiny species in Ecuador, measuring just 0.5cm (0.2in), is claiming a fresh bird dropping.

Not all **dung beetles** live in exotic locations like Africa. *Geotrupes vernalis*, in the related family **Geotrupidae**, is a common European species, busy burying dung from sheep and other animals in the warmer months.

Four different **dung beetle** species from Gorongosa National Park in Mozambique, an area rich in droppings from grazing mammals. Top left: *Garreta nitens*; top right: a tiny dung-covered pair showing that some species do cooperate in moving the ball; bottom left: *Kheper* sp., with huge shovel head used in digging; and bottom right: horned male of *Proagoderus tersidorsis*.

Stag Beetles — the family **Lucanidae** has always been a source of awe. Even before the truly exotic species from the tropics were discovered by naturalists, Europe already had its own giant stag species, performing ritual combat among the males using their huge jaws. These are male-only adaptations and not used to bite, but to overturn their opponents.

One of the jewels in the crown of the **stag beetle** family is *Phalacrognathus muelleri* from north-east Australia. It is impossible for a photo to capture all its glory, with multiple hues of bewildering depth changing as the beetle moves through light. This male is about 7.5cm (3in) long. Collectors used to spend a fortune for specimens of this rare species, but luckily now it is bred in captivity for this market.

Prosopocoilus torrensis is a **stag beetle** from northern Australia and New Guinea. It is smaller at 3.5cm (1.4in) and with less outrageous jaws than the Bornean species of the same genus.

Prosopocoilus mohnikei from Borneo is one of a group of large brown species in South-East Asia. This male is 6cm (2.4in) long.

An even rarer blue form of *Phalacrognathus muelleri* exists. It is so rare that a museum specimen is shown here.

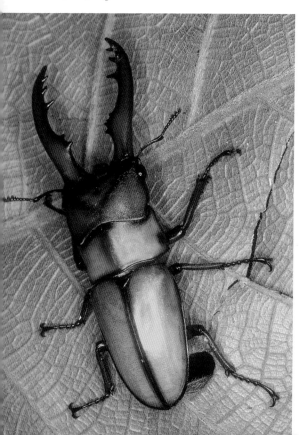

While this book showcases some of the more amazing world species of various insect groups, the above species, *Lissapterus pelorides*, is actually the best example of a typical **stag beetle**. From Australia, 2.5cm (1in) long.

Cyclommatus is a large genus of South-East Asian **stag beetles**. Most are shiny and even metallic, but this male of a species from Borneo is powder brown. 6cm (2.4in).

Lucanus cervus is the best-known **stag beetle** in the world, as it is one of the largest beetles in Europe, living in old oak forests. It uses its formidable antler-like jaws to fight with other males for access to females. The beetles strength is demonstrated not just by lifting the opponent, but by being able to do even when balancing only on two legs high up a tree, as shown here.

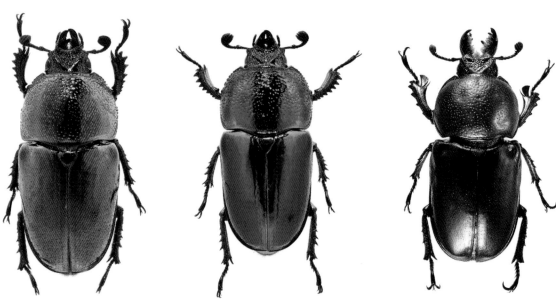

The genus *Lamprima* has some spectacular **stag beetle** species with great variations in size and shade. Top: *Lamprima adolphinae* is a fabulous example from Indonesia, 3.5cm (1.4in). Most *Lamprima* species have a metallic-green 'standard' form, but above are museum specimens showing the beautiful extremes within this group. From left to right: *Lamprima aurata* female, *Lamprima latreillii* female, *Lamprima varians* male, all from Australia and around 2cm (0.8in) long.

Jewel Beetles are another family of very noticeable, and historically collectible beetles. There are bout 15,000 species of **Buprestidae** in the world, which are often seen visiting flowers in the warmer months. They range in size from a mere 0.2cm (0.08in) to a huge 8cm (3.15in). Here a selection from this treasure trove.

Metaxymorpha gloriosa is from Australia, 5cm (2in) long.

Selagis (Curis) caloptera is a small but bright visitor to eucalypt blossoms in Australia, 1cm (0.4in).

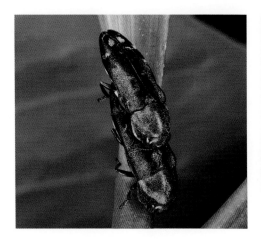

Curis sp. **jewel beetle** mating on *Acacia*, which is their host plant in Australia. 1.4cm (0.4in).

A green **jewel beetle** from New Guinea, measuring 4cm (1.6in).

Euchroma gigantea is the largest **jewel beetle**. It is from Central and South America and only has this yellow powdery sheen when young, before it rubs off. 8cm (3.2in).

Chrysochroa fulminans is a species in a widespread genus of South-East Asian **jewel beetles**. 5cm (2in).

This 1.8cm (0.7in) **jewel beetle** from Australia, *Castiarina letuipennis*, is quite unique as white is rare in the insect world.

Some African **jewel beetles** have a broad flat body, like this 2.5cm (1in) species from Madagascar.

Temognatha variabilis is typical of the large number of **jewel beetles** that are common on flowers in Australia. 3cm (1.2in).

Castiarina bucolica is a tiny **jewel beetle** from the flower-rich Western Australian coast. 0.8cm (0.3in).

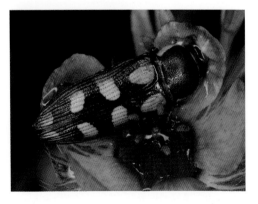

Another species in the genus *Castiarina*, which has hundreds of species across the continent of Australia. 1.4cm (0.6in).

Nascio simillima is an odd-one-out **jewel beetle**, with wood-like patterns rather than more normal garish attire of the family. From Australia, 1.5cm (0.6in).

At up to 6cm (2.4in), *Juludimorpha saundersii* is one of the largest **jewel beetles** in Australia, and one with a bizarre story. Beer bottles discarded along roads resemble females and attract males who try to mate with them. This phenomenon was decimating the species until the government introduced a bounty on the bottles and reduced the number of these temptations.

With more than 700 species, eucalypts in Australia represent a major source of food for insects when they flower. Here *Temognatha bruckii* is collecting sugars from these fresh blossoms.

This unidentified species of **jewel beetle** from the Western Australian flower belt epitomises the amazing patterns of this family. 2cm (0.8in).

Ladybird beetles or ladybugs, family Coccinellidae, are very recognisable, and very welcome by gardeners. The larvae and adults of the majority of species are voracious hunters of soft insects such as aphids, scale insects, mites and other garden pests. However, there are just a few species in this family of more than 6,000 species, which eat the foliage of some crops.

Ladybird larvae typically look like this long-legged, velvety-skinned beast, here devouring aphids, like most species do.

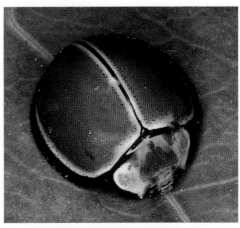

Not all ladybirds are spotted. This shiny species is *Antineda princeps* from the Australian tropics. 1cm (0.4in).

Harmonia conformis, the Large-spotted Ladybird, is very common and so renowned for its pest-eating prowess that it was exported from Australia to New Zealand to help protect crops there.

The Seven-spot Ladybird, *Coccinella septempunctata*, is common in Europe. This image was taken in Poland. It is a major aphid hunter, and so has been introduced to the USA to help fight the pests.

A new definition for shocking red – this very visible **ladybird** species is from savanna habitats in Madagascar.

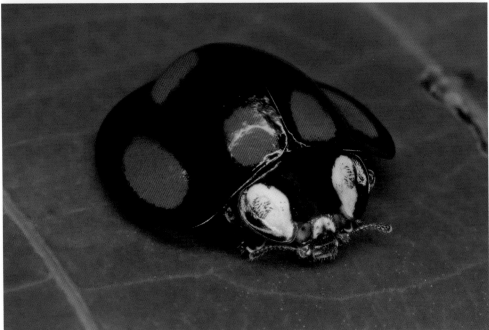

New Guinea is famous for many exotic insects, and among its many species is this **ladybird** in the genus *Australoneda*. It is large at 1.5cm (0.6in), with bold patterning and a flange extending the wing cases well beyond the body.

Many deep-sea creatures luminesce, and many fungi put on amazing glowing shows in wet forests at night. Among the insects, some midge larvae and three families of beetles have evolved the means to mix the chemicals which produce cold greenish light. The majority, more than 2,000 species, are in the worldwide family Fireflies, the **Lampyridae**; the Americas have a small number of species in the family **Phengodidae**, and also some click beetles, **Elateridae**, with glowing organs.

A typical **firefly** from Ecuador (or **lightning bug** as they are sometimes called in USA). As the sexes find each other by the lightshows the males produce, they have large night-vision eyes. The males have two segments of light-producing organs on the abdomen (as above), while the female usually only has one. Normally the males fly and produce the flashing, while the females sit still. A part of this family, the **Lampyridae** has females which mature without becoming beetle-shaped, looking like the hunting larva both sexes start as.

Atephylla sp. from Australia showing how from above **fireflies** look like any normal beetle, but on the underside the flashing organs set them apart. 1cm (0.4in).

In the Americas, the yellow-brown with pink pattern is a warning to predators that the insect is poisonous to eat. The **fireflies** are part of this sequence, which includes non-poisonous mimic species. Note the larva in this Costa Rican example has the same pattern.

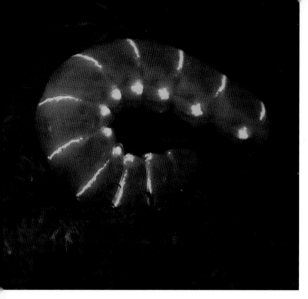

The beetles with the most, if not brightest, glow are the '**glow-worms**', also called '**railroad worms**', in the family **Phengodidae**. Unlike the fireflies, which blink their light organs constantly, these beetles, when larvae, glow all the time from many light organs. The females stay larvae-shaped after maturing and still glow, while the males become non-glowing, flying beetles. Only about 100 species are known, all from the Americas.

The largest **fireflies** in the world are in South-East Asia, in the genus *Lamprigera*. Larvae up to 11cm (4.4in) long hunt snails in the wet undergrowth of the rainforest. The females of this genus stay larviform at maturity and are sought by the flying beetle-bodied males.

The family **Rhipiphoridae** contains boat-shaped beetles with fan antennae, famous for many species having parasitic larvae that attach to bees and invade their larval cells.

The **click beetles** are a small but noteworthy family of beetles, the **Elateridae**. They possess a spring-loaded mechanism between the thorax and abdomen, which stores energy that can be released in a single loud 'click' which propels the beetle upwards. It is used as a startle defence, especially if the beetle is turned over onto its back by a predator.

This **click beetle** has evolved to blend in with tree bark, but is shown here on the wrong, contrasty, backdrop. It belongs to a large genus *Paracalais*, which contains big species in South-East Asia and Australia. 3cm (1.2in).

One of the brightest **click beetle** species is not from the exotic tropics, but from Europe. *Ampedus sanguineus* lives in spruce forests. 1.5cm (0.6in).

Most **click beetles** are brown, sometimes with serrated antennae (as above). They have the prominent division between the thorax and abdomen to allow a lot of flexing while using the click mechanism. *Megapenthes* sp., from Australia, 1.8cm (0.7in).

The genus *Semiotus* from Costa Rica contains brightly patterned **click beetle** species that are active during the day. 1.8cm (0.7in) long.

Another metallic **clerid beetle** – a species from the genus *Eleale*. Clerids spend a lot of their adult life on flowers.

The family **Cleridae** are elongate, often metallic and bright beetles with short clubbed antennae. Many are beneficial as hunters of timber pests. *Eleale* sp. from Australia, 1.5cm (0.6in).

African insects often depart from expected patterns for particular insect groups. This **clerid beetle** from the southern deserts shows original bold patterns. 0.8cm (0.3in).

Fan beetles are a small family, the **Rhipiceridae**, known for their elaborate fan-like antennae. *Rhipicera* sp., from Australia, 2cm (0.8in).

The family **Lycidae** has species all over the world which look much like this one (top) – brick-red with black extras. Because many species of this appearance are poisonous to eat, a long chain of mimics in other beetle and insect families has evolved, obtaining some protection from this general body plan. Below, clockwise from top left, are: a moth, family Oecophoridae; two jewel beetle species, family Buprestidae; and a beetle, family Oedemeridae. None of the mimics are poisonous.

In the highlands of New Guinea live several deep metallic **lycid beetle** species. They are probably also poisonous, like the ones above. 1.2cm (0.5in).

In Sri Lanka, many **lycid beetles** have this extra developed, serrated antennae. Larger surface area on the antennae often means a better sense of smell for detecting the pheromones of females.

The family **Melyridae** contains often bright beetles of many shapes. *Dicronalaius* sp. are often found on flowers in Australia, where they probably both predate and pollinate.

This bizarre creature is a larviform female **lycid beetle**, genus *Platerodrilus* from Malaysia. It develops as a **larva** this shape, but on the last moult to adulthood, does not metamorphose into a beetle body. It is 3.5cm (1.4in) long and feeds on microorganisms in wet rotting wood.

This deep red species of **lycid beetle** is from Thailand. The photo caught the moment of take off, which demonstrates the general beetle adaptation of hardened upper wings forming covers called elytra, and the larger soft flying wings, folded origami-like inside.

Balanophorus mastersi is a pollen-eating species in the family **Melyridae** from Australia. 1cm (0.4in).

A part of the **Melyridae** family has species with full-sized wings tucked into wing-covers only half as long as the abdomen. This metallic species is from New Guinea. 1.2cm (0.5in).

Fungus beetle, family **Erotylidae**. This Costa Rican species lives in the very wet, upland cloud forests. 1cm (0.4in).

One the most outrageously red beetles – a **fungus beetle** in the family **Erotylidae** from eastern Madagascar, 1.4cm (0.5in).

Thousands of species of beetles in many families are fungus-eaters. Shown here is an **Erotylidae** species from Costa Rica on bracket fungi.

The **Endomychidae** family has very distinctive larvae, which feed on fungi in the rainforests of Madagascar. Adult next page top right.

One of the brighter fungus beetles in the forests of Ecuador, from the genus *Erotylus*, family **Erotylidae**. 2cm (0.8in).

The **Endomychidae** family has rounded beetles with a flat flange extending the wing-covers. Their larvae (previous page bottom right) are wonderfully odd. From Costa Rica. 1.2cm (0.5in).

Last but not least of the **fungus beetle** families is **Languriidae**. They are shiny, elongate and often very strongly patterned, like this pair of *Anadastus albertsi*, from Australia, 1.4cm (0.5in).

South American forests can be very wet. Fungus is king here, and myriad insects have evolved to make use of it. The **Erotylidae** family has some of the most beautiful examples, like this species of *Gibbifer*. 1.5cm (0.6in).

The **oil beetles**, family **Meloidae**, have many showy and fascinating species. Another name for some species is '**Spanish Fly**', the insects whose deadly extract was a favourite assassination poison in the royal courts of Europe. The cantharidin chemical, in the 'oil' the beetles' exude, can cause nasty skin irritation, therefore their other common name is **blister beetles**. Their bold pattrerns serve as a warning to predators. Africa is where the most species reside.

Like most **oil beetles**, this member of the genus *Epicauta* has bold markings to warn predators to avoid its poisonous body. This adult from Thailand feeds on leaves, but its larva eat the hidden eggs of grasshoppers. 2.4cm (1in).

The **Violet Oil Beetle**, *Meloe violaceus*, is a common European species. Adults are wingless and tend to be on flowers eating pollen, but also laying thousands of eggs which hatch into parasitic larvae that specialise on native bees. Back in the bee nests the larvae eat their eggs and baby food supply. 3cm (1.2in).

The extreme flower diversity in southern Africa has allowed some beetle families to flourish more than on other continents. **Oil beetles,** such as this *Ceroctis* sp., abound. 1.5cm (0.6in).

This plain beetle has no warning markings but is still an **oil beetle**, and one well known for blistering skin in contact. *Epicauta* sp. from Mozambique, 1.2cm (0.5in) long.

Mylabris oculata is one of the larger African **oil beetles**, common on flowers in early spring. 2.6cm (1in).

The family **Oedemeridae** is related to oil beetles and members often share the bright dress, although not the poisonous body. *Oedemera femorata* from Europe. 1.8cm (0.7in).

Above the bee's legs, note late instar parasitic **larvae of an oil beetle**, about to be transported to the bee's nest.

Nuttall's Blister Beetle, *Lytta nuttalli*, is perhaps the most glorious North American beetle. However it is also a pest because it feeds on rapeseed and of course causes blisters if handled.

The family **Tenebrionidae**, with more than 20,000 world species, is quite varied. Many are called darkling beetles, but most have no common names. The majority are dark, mostly nocturnal beetles associated with rotting wood, detritus and fungi. Some are eye-catching daytime flower-dwellers, while others are pests of stored products. A large part of the family are desert specialists surviving in extremely dry places where rain may not happen at all in their lifetimes. The Namib Desert in south-west Africa is a key location for these super-adapted species. The first images here are a showcase for their forms and habits.

White is rare among insects. Even though it would seem prudent to be reflective white in the heat of a desert floor, *Cauricara eburnea* is one of only a few white species in the Namib. Long legs keep it off the hot sand. 1.4cm (0.5in) body.

Onymacris plana is one of the flattest Namib beetles. Long legs raise it above the hot sand, and the flat body can easily burrow into cooler sand. Body 2cm (0.8in) across.

Calosis amabilis is another slippery Namib species that both runs fast and can burrow fast. 1.8cm (0.6in).

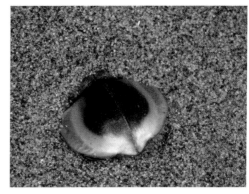

Lepidochora discoidalis perfectly illustrates the excellent survival adaptations of the Namib beetles. Its slippery body can swim through sand in daytime to escape the heat and search for detritus matter to eat. The Namib rarely sees rain but the cold Atlantic meeting the hot sand causes dense night fogs to roll in deep inland. *Lepidochora* beetles dig a furrow on the ocean side of a dune top (below left) and drink the water that condenses on the raised edges. 1.4cm (0.5in).

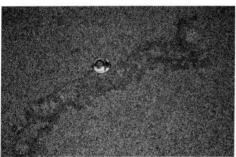

Onymacris unguicularis is another fog-drinking Namib beetle. It stands on dune tops at night with body raised, and lets condensed water trickle down to its mouth. 2cm (0.8in).

The shield **darkling beetle**, *Cardiosis fairmairei*, is a slippery species, easily swimming inside the dunes during the day. The bold design has been used by local peoples. 1cm (0.4in).

American arid zones have many black **darkling beetles** such as this *Coelocnemis* sp. from New Mexico. The weird posture is a common darkling beetle defence pose, a bit like a rearing spider.

Australian arid zones also have many **darkling beetle** species known as **pie-dish beetles**. They are flat, sometimes wingless, and live under bark if trees are present. This species in the genus *Pterohealeus* measures 1.8cm (0.7in).

New Guinea again surprises with wonderful exceptions to the rule. Here a very bright **darkling beetle** from the highland rainforests, which is also day active to show off its finery. 2.8cm (1.1in).

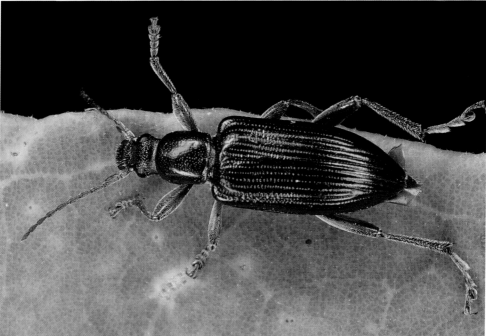

A family called **Alleculidae** was added to the **darkling beetles,** resulting in the addition of a number of brightly marked flower visitors into the mix. Shown here are two species of *Aethyssius* from Australia. 1cm (0.4in).

A group of **darkling beetle** species in South Africa have the novel habit of tapping their feet in species-specific rhythms to attract a mate. *Moluris pseudonitida*, 1.8cm (0.7in).

The wet forests of Borneo teem with rotting wood- and fungus-eaters, and this large **darkling beetle**, *Robustocamaria* sp., is commonly seen at night. 3cm (1.2in).

Cyphaleus planus is part of a genus of metallic **darkling beetles**, found from South-East Asia to Australia. 2.5cm (1in).

On African beaches, under high-tide algal mats, live these **darkling beetles** which feed on dead algae. *Pachyphaleria capensis*, 0.8cm (0.3in).

Members of the **Lagriinae** subfamily of the **darkling beetles** have a characteristic narrow thorax, like a long 'neck'. They are conspicuous by being day-active. From New Guinea, 1cm (0.4in).

This oddly elongate and narrow-waisted **darkling beetle** is from Madagascar, where many insect groups have 'exceptions to the rule' species. 2cm (0.8in).

These distinctive beetles have several common names, including pin-tail beetles, tumbling flower bugs and fish beetles. They are very fidgety, fast and quite slippery to hold, spending their adult life in flowers. This species in the genus *Hoshihananomia* is in the family **Mordellidae**. 1cm (0.4in).

Like the Mordellidae family, the **Rhipiphoridae** have pointed slippery shapes. Adults are found on flowers, but the larvae are parasites of beetles, cockroaches and wasps. In some the females remain in a wingless larviform shape even when mature. *Macrosiagon* sp. from Australia, 1cm (0.4in).

A small **fish beetle**, family **Mordellidae**, from Borneo. 0.8cm (0.3in).

One of the largest families of beetles are the leaf beetles, **Chrysomelidae**. As the name implies, they are plant eaters, both as adults and larvae, and as such have a huge potential food supply to adapt to. More than 32,500 species have been named so far, in many diverse subfamilies. From minute 'flea' beetles as small as 0.12cm (0.05in), to large metallic 'frog-legged' beetles up to 5cm (2in) long. An introduction to these herbivores follows.

Starting with the big species, the subfamily **Sagrinae** contains beetles distinguished by the large rear legs of the males – the source of the common name **'frog-legged' beetles**. The genus *Sagra* contains the giant and spectacular species, mainly from South-East Asia. Both have a body-only length of about 3.5cm (1.4in).

The most numerous subfamily of **leaf beetles** is the **Chrysomelinae**, with many rounded, dome-shaped beetles, personified by the hundreds of species of the Australian genus *Paropsis*. Most are about 1cm (0.4in) long, with an infinite variety of patterns and textures, from shiny to dusty. The fat larva (top right) lives near the adults.

Among so many leaf-eaters there are a few pests. The famous **Colorado Potato Beetle**, *Leptinotarsa decemlineata*, from the USA is among them. 1.2cm (0.5in).

Spines are a feature of many species of the **Hispinae** subfamily of **leaf beetles**. The adult ones are sharp, but their larvae are adorned with softer ones. This species is from Borneo, 1cm (0.4in).

Two very pretty **leaf beetles** from the jungles of Costa Rica. On the left is *Disonycha* sp., 1.4cm (0.5in). On the right is *Calligrapha fulvipes*, 1cm (0.4in).

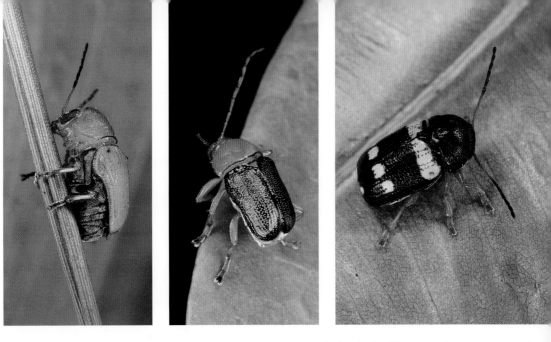

The **Cryptocephalinae** subfamily of leaf beetles has these more elongate, robust-looking beetles. Most are small to very small, and are generally found singly on leaves. Shown here are three Australian species from the genus *Aporocera*, all about 0.8cm (0.3in) long.

There is red and then there is RED. These two tropical **leaf beetles** push the definition to new limits. On the left an un-identified but not missed species from Ghana, 1.2cm (0.4in), and on the right *Lilioceris cheni* from Thailand, 1cm (0.4in).

One of the more common European **leaf beetles** is this metallic blue species, *Oreina* sp., here in an oak forest in Poland. 1.2cm (0.5in).

Parental care is rare among beetles. Here *Tinosis maculata* from Australia is protecting its brood, also aided by them covering themselves in dirt and their own droppings. Adult 1cm (0.4in).

One offshoot of the leaf beetles are the aptly named **tortoise beetles**, subfamily **Cassidinae**. All have extended and flattened wing cases, which hide the body and legs fully from above. Sometimes the flange is see-through, and almost always the markings and patterns are bold and original. The next two pages present entertaining examples.

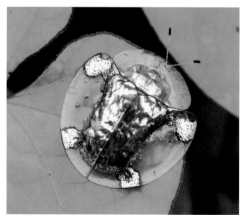

These bold markings belong to *Stolas punicea* from Belize. 1.2cm (0.5in).

This unidentified **tortoise beetle** excels in a mix of garish and subtle, with pure gold amid silver and sky blue. 1cm (0.4in).

Above is a very green, unidentified species of **tortoise beetle** from the rainforests of Madagascar in a rare upright pose. Below is a truly weird species of **tortoise beetle**, also from Madagascar, with extra horns and in a more normal flattened protective pose. Both about 1.2cm (0.5in).

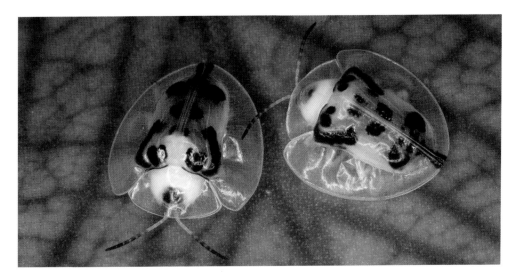

Tortoise beetles in the Australasian genus *Emdenia* have a fully see-through flange, protecting yet showing their legs. 1.2cm (0.5in).

The **larvae of tortoise beetles** maintain the weirdness, adding cast skins to the already spiny body to hide their general appearance.

A bull's-eye pattern on an edible insect is a bit unfortunate. *Ischnocodia annulus* from Costa Rica. 1cm (0.4in).

Another unknown species of **tortoise beetle** from Madagascar, metallic, gold and see-through. 1.2cm (0.5in).

Opposite: *Batocera* is a genus of about 60 species of large ornate **longicorn beetles**, found from tropical Australia to Africa, but not in the Americas. The broad pads on their feet and the robust long antennae are typical of the family. This individual is from New Guinea. 7cm (2.8in).

Longicorns, or long-horned beetles, family **Cerambycidae**, are another of the iconic, huge beetle families, historically popular with collectors for their often fabulous forms. The name suggests they have long 'horns', however unlike beetles that do have horns, the name here refers to long and very long antennae, up to several times the body length. Longicorns are found all over the world, but the majority of the 25,000 species are tropical. Their larvae tunnel in living and dead wood.

Batocera boisduvali is from Australia and New Guinea. Like many species in the genus, its larvae live in fig trees. The close-up of face (below) shows their very strong jaws, which are used to carve a niche into the tree to lay eggs. 6cm (2.4in).

Not to be outdone by the profusion of tropical **longicorn** species, Europe has some very beautiful examples. Above left is *Leptura maculata*, to the right is *Stictoleptura cordigera*, and below is an iconic favourite species, *Rosalia alpina*, which features on the postage stamps of at least six different countries.

The genus *Clytus* is found almost all over the world. The first named of these mild wasp mimics is *Clytus arietis* from Europe. 1.4cm (0.5in).

Dorcadion fuliginator belongs to a group of **longicorns** which live on the ground rather than in trees. From Europe, 2cm (1.8in).

A splendid unidentified **longicorn** from the eastern rainforest of Madagascar, 2.5cm (1in).

One of the beetles that reminds us how exotic the South American insect fauna is. *Acrocinus longimanus* has these spectacular males with, as the name says, 'long-hands'. The body is 7cm (2.8in) long, but the hand-span is 20cm (8in) wide.

Platymopsis viridescens is part of genus whose larvae specialise by living in acacia plants in Australia. 2cm (0.8in) long.

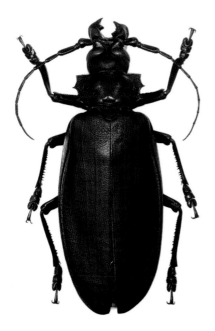

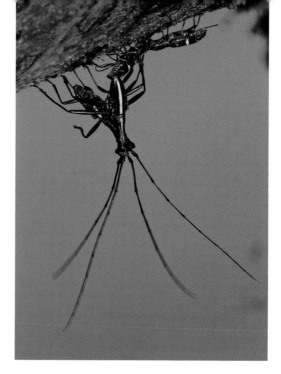

This museum specimen belongs to *Titanus giganteus*, the largest longicorn and one the largest insects in the world. It is a rare find in South America, at up to 17cm (6.8in).

High drama in the jungles of Indonesia. **Longicorns** of the genus *Gnoma* assemble for mating rituals under a log. Two males fight for the smaller females watching on. Males 2.5cm (1in).

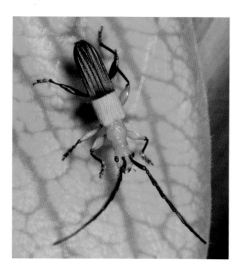

Tropocalymma dimidiata is a frequent flower visitor on the east coast of Australia. 1.2cm (0.5in) long.

This spectacular **longicorn beetle** is found in Madagascar, where they sometimes congregate in sunny clearings in the rainforest to show off their shimmering wares. 2cm (0.8in) long.

Weevils are the most speciated animals on the planet. Something about the adaptations evolved by these long 'nosed' generally wood-, root- and leaf-feeders, has resulted in more than 62,500 named species across seven families. The true weevil family, the **Curculionidae**, has about 55,000 of these. Weevil mouthparts are mounted on an elongate rostrum, and in some sopecies this is even longer than the body. Apart from feeding, it is used to dig fine holes in plants to deposit eggs. Most species are rather dull, a few are famous pests of stored products (although most do not interact with humans) and some are spectacular exotic creatures. Sizes vary from about 0.1–6cm (0.04–2.4in).

Not a typical brown weevil. *Eulophus* sp. in the highlands of New Guinea are mainly blue and are day active, displaying their beauty. 2.5cm (1in).

The Indonesian **'diamond' weevil** from Sumatra. 1.8cm (0.7in).

One subfamily of weevils specialises in palms. They arrive in numbers where palm trees have been damaged and weep sap. Many are very large, like this Australian species, *Iphthimorhinus australasiae*. 3.5cm (1.4in).

Left: black **clown weevil**, *Pantorytes stanleyanus*, from Australia is part of a series of tropical Asian species with bright dots and dashes. Right: a relative from New Guinea. Both about 1.2cm (0.5in).

Peridinetus cretaceus from Costa Rica lives in the wet cloud forests at high altitude. 1.5cm (0.6in).

A minute **jeweled weevil** from Costa Rica, on *Heliconia* flowers. 0.5cm (0.2in) long.

In every insect family the first genus to be named takes the family name. Thus the first weevils, **Curculionidae**, were given the genus name *Curculio*. These are tiny, very long 'nosed' weevils known as acorn and nut weevils as they use the jaws at the end of the rostrum to dig deep holes in nuts and lay eggs inside. Left: *Curculio betulae* from Europe; right: a *Curculio* species from the USA. Both about 0.8cm (0.3in).

Within the weevil group is the family **Attelabidae**. The more than 2,000 species of these often bright snub-nosed weevils are leaf-curlers. After mating the female expertly cuts a leaf in such a way that she can curl it into a tight whorl, into which she lays an egg. The larva is protected by eating the leaf from inside. *Euops* sp. from Australia, 0.6cm (0.2in).

The family **Attelabidae** has some truly strange species. A group of them are known as **giraffe weevils,** and this species, *Trachelophorus giraffa*, is the most striking of them all. It is found in Madagascar, and used as an example in many publications to show just how weird the fauna of that fascinating country can get. Only males have the super long neck. 2cm (0.7in).

Europe also has some showy leaf curlers from the family **Attelabidae**. Shown here is the **Hazel Leaf Curler**, *Apoderus coryii*. 0.8cm (0.3in).

This metallic member of the **Attelabidae** demonstrates another show-off method within this family. From New Guinea, 0.8cm (0.3in).

Weevils, more so than other beetle families, have a number of species adapted to mimic bird droppings. Not a very edible morsel for predators looking for beetles to eat. On the left is an extra clever disguise from Ecuador, using the hole in the leaf for additional realism. On the right a nice general bird-dropping look, from Ghana.

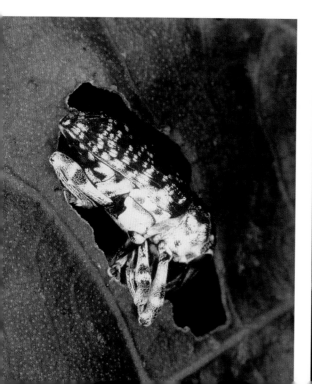

When Australia was first explored in the 18th century, botanist Joseph Banks was so astounded by the variety of new plants in the bay of first landing that he called it Botany Bay. Among these plants was this beautiful weevil, which was promptly named the **Botany Bay Diamond Weevil**, *Chrysolopus spectabilis*. 1.8cm (0.6in).

An unidentified species of tiny **weevil** from New Guinea, 0.5cm (0.2in).

Some **weevils** hang onto branches in quite bizarre ways, until you consider that such poses can be a deception based on odd burrs and knobs on tree branches. From the deserts of Australia, 0.6cm (0.2in).

A genus of weevils in New Guinea has the common name of **spider weevils**. They live on tree trunks and branches and resemble large tarantula-style spiders, even down to the way they walk slowly with deliberate spider-like leg movements. Both *Arachnobas* sp., and about 2cm (0.8in) body size.

Some insects are just very expressive. *Hoplocopturus* sp. from South America perform complex mating rituals at dazzling speed, now and then stopping in this comic position.

This well defended spiny **weevil**, *Catasarcus carbo*, lives on the ground in the western deserts of Australia, where extra protection may well be needed. 1cm (0.4in).

This amazingly streamlined **weevil** from New Guinea would be a slippery prey item for some predators. The head is up. 1cm (0.4in) long.

An unidentified weevil in the **palm weevil** group, from Thailand. 2.5cm (1in).

This **weevil**, probably *Curculio* sp., from Malaysia, has one of the longest noses in the business, although not the longest. A similar species has one three times its body length. 2.5cm (1in) with nose.

Weevils always seem to have an attitude. This large long-nosed and tusked *Glochinorhinus evanidus* from the northern Australian rainforest did not budge when facing the camera. 3cm (1.2in).

Blue beetles are rather rare. This *Eulophus* sp. is one of many blue **weevils** from the New Guinea Highlands. 2cm (0.8in) long.

On the left is a tiny unnamed **weevil** standing on its head, and on the right is a jumping spider eating a mite. Both live in the same forests in New Guinea, and the theory is that they are mimicking each other. Who gains more protection is an interesting question. The weevil performs the headstand in order to use its longer back legs to imitate the threat pose of spiders. The spider is *Coccorchestes ferreus*. Both 0.4cm (0.15in).

The 3,000 species of the weevil family **Anthribidae** are known as **fungus weevils**. The males (left) sometimes have very long antennae, and the rostrum, or nose, is usually short and wide. Larvae are associated with fungus-ridden logs. From New Guinea, both 1.4cm (0.5in).

This New Guinea **weevil** is inserting its 'rostrum', the tubular mouthparts, into a fruit that is still green. It will find a way into the seed and then lay an egg or more into the breach, for the larvae to eat the seed. 1cm (0.4in) long.

The last but not least of the **weevil** families is **Brentidae**. About 4,000 species around the world are mainly associated with freshly felled wood. They are usually defined by a very skinny elongate body, with the rostrum up to as long as the body. Males are often more ornate than females. This long species is from Costa Rica. 2.5cm (1in) long.

A male **brentid beetle** striking a pose. *Ectocemus decimmaculatus* is found from Indonesia to Australia. Its larvae live in dead wood tunnels. 2cm (0.8in).

This round-bodied **brentid beetle** is getting its fill of this fruit in a Madagascar rainforest.

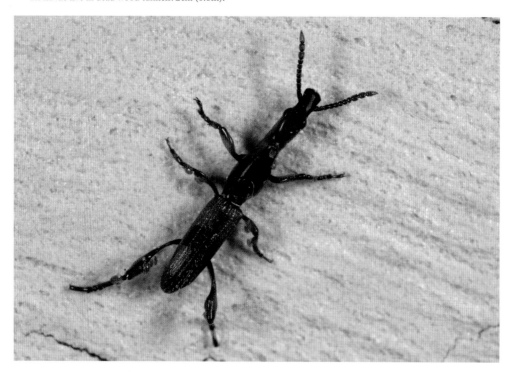

Above and opposite: Members of one part of the **Brentidae** family have ornate rear legs and a stubby rostrum, or no rostrum. They live in tunnels made by ambrosia beetles, also a type of weevil. Opposite: *Calodromus* sp. from South-East Asia, 1.2cm (0.5in) body. Above: *Cyphagogus delicatus* from Australia, 0.4cm (0.15in).

STYLOPIDS

Order STREPSIPTERA

600 species in 9 families

This tiny order of tiny insects is like no other. They range only from 0.5–4mm (0.02–0.16in). All species are internal parasites of other insects. Females, in all but one family, never lose their grub-like form, even after undergoing metamorphoses to adulthood. Males develop into flying insects, although the front wings are reduced to little 'clubs', and the back wings are very large and oddly square. True flies, the only other insects with a single pair of wings, have reduced their hind-wings to 'clubs' instead. The males therefore fly around, seeking out the pheromones produced by the females, who do not leave the insect host they parasitise. After pupation their head and part of the abdomen sticks out through a join on the host's abdominal segments. The male finds her and mates, after which she produces from hundreds to a thousand eggs, which grow and mature inside her body. What hatches out are minute free-living first-stage larvae called 'triungulins'. These disperse and wait to be picked up by new hosts to start the parasitic cycle again.

The different families of stylopids tend to always parasitise the same insect groups. Various planthoppers, grasshoppers, bees and wasps are among the most targeted.

The weird winged male **stylopid**. Body 0.3cm (0.12in). Despite the suggestively dangerous appearance of some species, scorpionflies do not bite, and neither do they interact with humans.

A bee in the family Melittidae that has been parasitised by a wingless female **stylopid**. She is under the winged male mating with her. Later she will disperse hundreds of minute larvae that will search for more hosts to parasitise.

This paper wasp, *Polistes dominulum*, has been parasitised by a **stylopid** species, *Xenos vesparum*. The wingless female, which grew inside the wasp, is protruding through an abdominal segment join (highlighted by the arrow), waiting for a winged male to mate with her.

SCORPIONFLIES

Order MECOPTERA

500 species in 9 families

A formidable name for a small harmless order of insects. The name derives from the males of one family, the **Panorpidae**, having a habit of holding their complex last segment curled upwards like a scorpion's tail. These insects are related to the lacewings, and the family is nearly as old, being one of the first to have a full metamorphosis life sequence with a larvae and pupa before the adult stage. The larvae are caterpillar-like grubs, living near the soil and feeding on living and dead plant matter. The adults are mostly large-sized, and many of them are hunters. Their heads have the jaws mounted on a protrusion of the other mouthparts, a rostrum. About half the species hunt smaller insects, either on the wing or by stealth. The family **Bittacidae** are known as hanging flies, as they hang from vegetation by their long forelegs, and use very modified rear legs to snare passing prey. The fourth and fifth segment of their tarsi, or 'feet', are modified as serrated ends of pincers which capture flying insects.

Prey capture is crucial to mating in the hanging flies. Only males with prey presented to the females as a nuptial gift attract a willing female. This is a clever adaptation which gives the females the protein they need to make eggs after mating. Eggs of most species are laid on the ground, where the larvae live either on the surface or in burrows. After seven or more moults, these pupate into adults.

The most bizarre group of scorpionflies are the snowflies, found mainly in the Northern Hemisphere. They are small, with wings reduced to stubs, and are active on snowfields. There they seek dead insects and other detritus, immune to often subzero temperatures.

The classic **Common Scorpionfly**, *Panorpa communis*, family **Panorpidae**, from Europe, demonstrating where the common name comes from. The upturned sting-like back end, is actually a complex mating apparatus of the males, but the resemblance is uncanny. 2.2cm (0.9in).

There is a story here. The '**hanging flies**' are a family of scorpionflies, the **Bittacidae**. They spend much time hanging by their front legs, with specialised grappling rear legs hanging down ready to catch prey. At mating time, the males need to catch a gift morsel for the females. They then advertise this event with pheromones released by vesicles, the reddish blobs distending on its body. The female receives the gift only after mating, and has guaranteed protein for egg-making. Entomologists have observed that the size of prey offered changes the mating time. Tiny insects might result in 10 minutes. A housefly-sized insect results in up to half an hour, but this male has caught a huge crane fly, easily four to five times the size of a house fly. This is a species from the genus *Harpobittacus* from Western Australia, body 2.5cm (1in) long.

There are several insect species famous for being active on snow. The '**snow-flea**' is actually a small stubby scorpionfly which feeds on mosses, even through the winter. The reduced curled wings cannot fly, but help the male hold the female during mating. This species is *Boreus hiemalis*, family **Boreidae**.

A **scorpionfly** from Thailand showing very clearly where its name came from. In some species males have these very ornate last body segments.

FLEAS

Order SIPHONAPTERA

2,500 species in 16 families

Fleas are much maligned, but as a machine they are engineering marvels. The body is flattened sideways and made of extremely tough outer skeleton segments which overlap backwards, making them very slippery in one direction. The important joins are protected by rows of flat spines, called combs, which act to lessen damage from a scratching host, and also as anchor points to not be dislodged by the scratching. Their rear leg, and what would be flight muscles, are composed of resilin – an elastic protein so efficient that it can release 97 per cent of energy stored within. Before jumping the muscles in the thorax compress three different segment plates, which have locking mechanisms to hold the pressure. When ready to jump, the legs are raised off the surface, the thorax energy is released with an audible click, and then the legs push off to add a second wave of energy. Combined, the acceleration rises up to a staggering 140 gravities, propelling the flea too fast to see, and up to about 35cm (14in), or 85 body lengths upwards.

One more engineering oddity is that their mouthparts, fused into thin stylets, are driven into the host's skin by the action of the same resilin system as jumping. In this case releasing the tension in the resilin causes a hammer-like action of piercing.

Fleas vary from about 0.1–0.9cm (0.04–0.33in), although most are less than 0.5cm (0.2in). All are ectoparasitic, meaning that they are parasites on the outside of the host's body. Unlike the lice, which have more bird parasite species, fleas are mostly mammal parasites. They would have evolved from insects which lived in the nests of mammals, and many species still do so. But the majority live on the host body for most of their adult life. Eggs are generally thrown off the host, and hatch in the nest, and sometimes more freely. The larva are tiny grubs which feed on the 'scraps' such as the dried, blood-rich faeces of their parents. Some have the rather fetching habit of 'begging' for food from adult fleas, by tugging on their bristles, and sometimes being rewarded with regurgitated blood.

And now for the nasty part. Fleas are famous for being the carrier of bubonic plague, the 'black death' of medieval times. Only one species is the culprit, the rat flea, which is the vector between humans and rats in cities. Plague is still around, oddly even in America, but modern city hygiene practices make it very rare in humans, and modern drugs can treat it effectively.

Dog Flea, *Ctenocephalides canis*, on the white fur of a dog.

Cat flea, *Ctenocephalides felis*, on human skin.

FLIES

Order DIPTERA

160,000 species in 172 families

Flies are a very distinct group of insects due to having only one pair of wings. In ancient Greek, Diptera literally means 'two wings'. The second pair is modified into club-like 'halteres' which serve as flight stabilisers, allowing for the sort of acrobatics house flies perform to escape our swats. Other flies use this gyroscope-like effect to hover perfectly still, or suddenly fully reverse direction. Like the hover flies.

Their mouthparts are also unique, based on a kind of complex mop which is used for dabbing at largely wet food. However in some groups these mouthparts have adapted for piercing and sucking – with short rough stylets as in the horse flies and robber flies, or with long tubes as in the mosquitoes. It is only the females in these groups which suck blood from vertebrates, as they use the protein to make eggs more quickly.

The ability of some the piercing and sucking flies to deliver serious disease to humans, and the general peskiness of the house flies, gives an unfair reputation to the majority. Most of the 172 families of flies do not interact with us directly, but perform the very important roles of pollination and nutrient recycling. We may not like the house fly and bush fly adults, but their larvae dispose of and recycle most of the dead animals, and the droppings of the living.

Fly larva tend to be very simple grubs, called maggots, in most terrestrial families. The more primitive fly families often have aquatic larvae. The mosquitoes, midges and gnats, have larvae with names like wrigglers and blood worms. Some, like the mosquitoes,

did not evolve gills, but come to the surface to breathe through a siphon. Others are so small and soft that they can get oxygen from the water directly through their skin, and some have small gills attached to their breathing holes, the trachea. Pupating under water can be a problem for the emerging dry-land adult stage. The mosquito pupa floats on the surface like a boat as it emerges, while others store a bubble of air which lets the emerging adult bob to the surface.

There are roughly two subgroups of flies. The primitive Nematocera contains delicate elongate species, with long legs and long, sometimes frilly antennae. The mosquitoes, midges and gnats and more.

The Brachycera are more advanced, more robust, mainly terrestrial flies with short non-frilly antennae. In evolutionary terms, the most advanced flies are the more 'normal'-shaped families like the house flies.

Scavengers make up a large part of the order. Not just dead animals are recycled, but anything rotting and organic. Flies never evolved mouthparts for dealing with direct living plant feeding, but some have adapted to leaf-mining, feeding on the very soft tissues inside. Others compel plants to make homes of soft tissue for their larvae. They chemically induce the plant to form galls, which also hide the maggots from the outside world.

Many families are hunters as adults and some as larvae. The robber flies, **Asilidae**, are agile, sometimes large hunters, catching prey on the wing like dragonflies. Long-legged flies, shore flies and others tend to wait for prey come by. The other form of predation flies are famous for is parasitism. Some families, especially the **Tachinidae**, parasitise the larvae and pupa of many insects, including beetles and butterflies. The more highly adapted parasites include the louseflies, botflies and keds, which feed on vertebrate flesh, and wingless protective-plated flies like the bat flies.

The following is a tour of the flies, starting with the more primitive groups in the Nematocera. These families tend to be delicate, elongate, with long-antennaed adults and often aquatic larvae. The most famous are the mosquitoes, the Culicidae, and the many midge families.

A good demonstration of what flies do so well, is hovering. Here a **bee fly** from South Africa surveys its feeding territory and mating potential while hovering for hours on end.

Mosquitoes have also adapted their mouthparts into a long needle-like tube. Often this is very thin, and the use of a numbing chemical makes the bites almost unnoticeable. This hairy species is one of the largest in the world, *Aedes alterans*, the **Scotch Grey** from Australia. It attacks in numbers and its fat 'needle' has a potent sting.

The **dengue mosquito**, *Aedes aegypti*, is one of the most studied, and most dangerous insects in the world. Dengue, Zika, Chikungunya and many more diseases can be its 'gift'. Note that the actual needle of the 'syringe' is minute (the thin red line just below the head). Most of the apparatus is a protective sheath, moved aside here (see arrow).

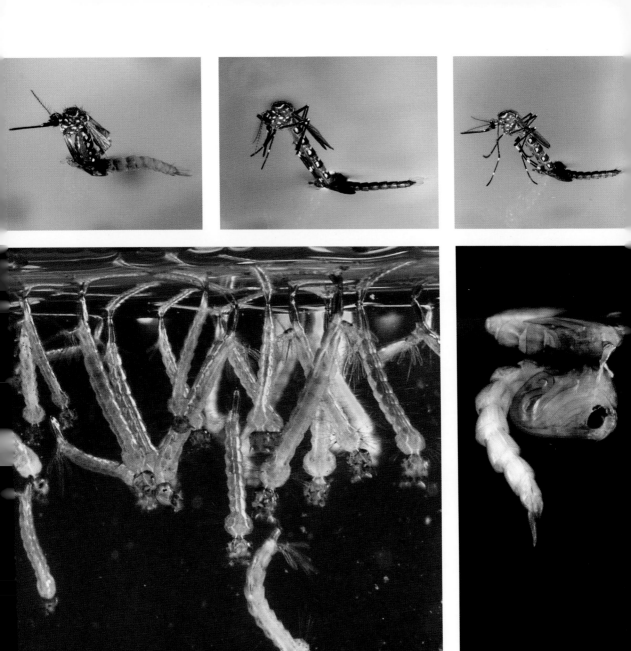

Mosquitoes have aquatic larvae called wrigglers (above left), and even aquatic free-swimming pupa (above right). Therefore the transition from water to a dry flying adult needs care. The pupa comes to the surface and floats like a little boat as the new adult very slowly emerges upwards. When most of the way out (top middle), it balances impossibly above the surface, and then the legs pull fully free and it slowly stands on the water surface while its new wings dry and harden prior to take off. This is the **dengue mosquito**, *Aedes aegypti,* and the pupal skin in all pictures on the top line, is just under the surface of the clear water.

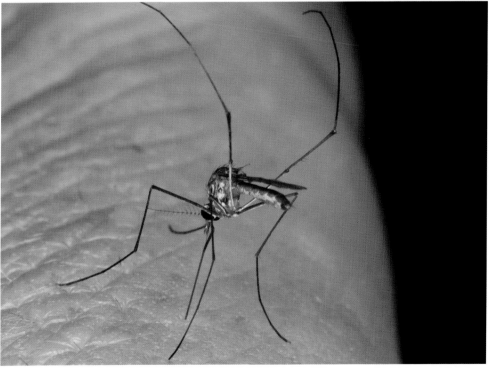

There is a general belief that **mosquitoes** only bite warm-blooded creatures, but as they find their prey by the carbon dioxide it exhales temperature is not a limitation. *Uranotaenia* species are frog-specialist mosquitoes. The individual at the top is in Australia, while above is another often cold-blooded biter, *Tripteroides* sp., in New Guinea, only resting on the photographer's finger. 0.3cm (0.12in) and 0.4cm (0.16in) bodies.

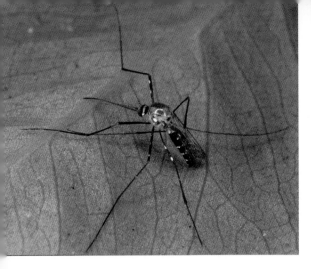

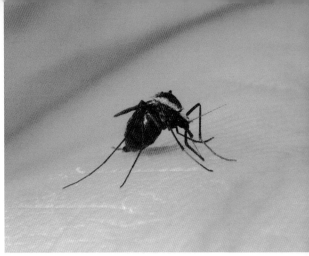

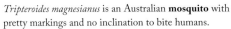
Tripteroides magnesianus is an Australian **mosquito** with pretty markings and no inclination to bite humans.

Below: One of the larger **mosquitoes** to bite humans, *Aedes vittiger*, often attacks in numbers, and their fatter proboscis actually stings going in. From Australia, 0.9cm (0.36in).

If uninterrupted in the process of blood-feeding, **mosquitoes** can drink in two to three times their body weight. They then fly very slowly away to a hiding place to digest the load and make eggs. *Aedes lineatopennis*, Australia.

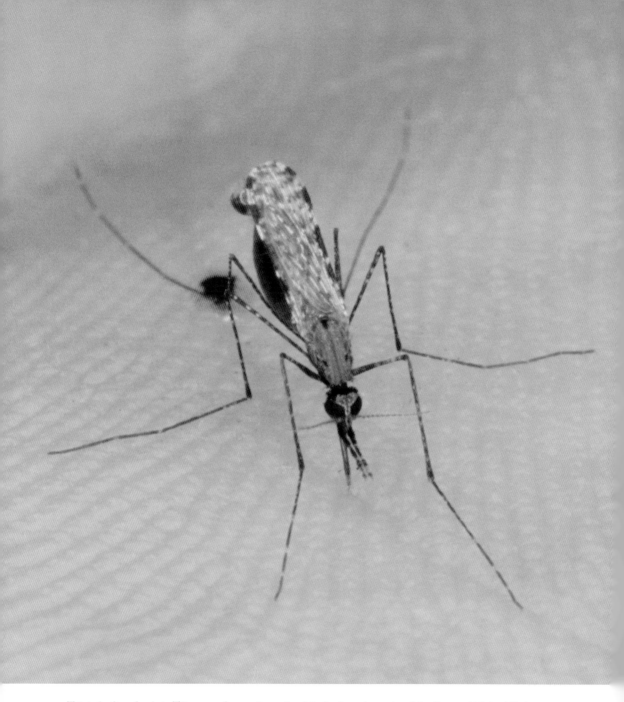

This is the face of malaria. This genus of **mosquitoes**, *Anopheles*, has been the carrier of the disease which has killed more humans over time than any other method, including wars. As the female drinks, she is filtering the red cells from the plasma and ejecting the less useful plasma on the go. About 30 tropical species can infect humans, but more bite birds and other animals. 0.8cm (0.3in).

Aedes albopictus is not yet the number-one enemy **mosquito**, but it is a very tenacious invader species spreading round the world at alarming speed. It is a carrier of the Dengue group of diseases, which also includes Zika and Chikungunya. With warmer winters, it is now established in parts of Europe and the USA, and poised to deliver epidemics.

The largest **mosquitoes** in the world are not human biters, not even blood feeders, and are beneficial insects. The metallic, curved-beak species of *Toxorhynchites* have huge larvae which eat other mosquito larvae. They have been used in dengue mosquito control, especially in tyres, where each of their larvae eat 50 or more larvae of the dangerous species every day. Adult 1.4cm (0.5in).

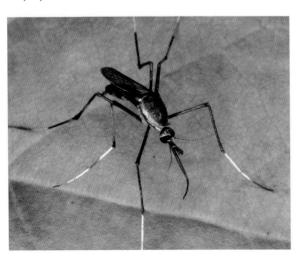

Non-biting midges, 8,000 species in the family **Chironomidae**, are among the most abundant organisms in freshwater habitats. Their larvae are aquatic, where they may be half of all animal species, filter feeding, leaf mining and mud dwelling. They do not bite, and are instantly differentiated from the similar-sized mosquitoes by lacking a biting proboscis. The males often have long front legs and are often green. Bloodworms, tiny larvae filled with haemoglobin to live in oxygen poor muddy habitats, are also Chironomids (right).

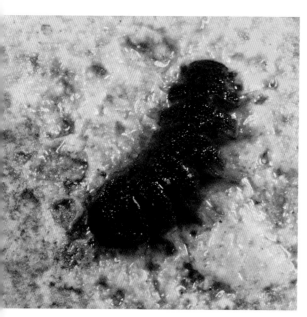

Torrent midges, Blephariceridae, have adapted to the extreme habitat of river rapids and waterfalls. Their larvae attach to the smooth rock via many sucker pads, facing upstream and filtering micro-organisms and detritus from the deluge.

The 15,000 species of **crane flies**, also called 'daddy-long-legs,' are in the family **Tupulidae**. They are among the oldest fly lineages, dating back 250 million years. Larvae of most species live in water or moist nutrient-rich habitats such as rotting vegetation. *Nephrotoma* species are found around the world, usually with these yellow-and-black patterns. 2cm (0.8in).

Crane flies are often encountered mating. The point of this image is to show that one subfamily of **Tipulidae** has thin stiletto 'beaks'. This further confuses them with mosquitoes. However crane fly legs are longer, and none bite – the beak is to stab into deep flowers for nectar.

This ethereal opalescent **crane fly** lives in the wet, mossy, cold cloud forests of Costa Rica, at 3,000m (10,000ft) altitude. 2cm (0.8in).

The **Cecidomyiidae** are the **gall midges**. Their larvae chemically force a plant to produce a tumour-like growth known as a gall, making a safe edible home. Some adults have the bizarre habit of sitting in a spider web and wobbling it about. The spiders seem not to mind. The 6,000 named species are probably only a fraction of the whole, as most galls are not yet linked with the tiny elusive adults.

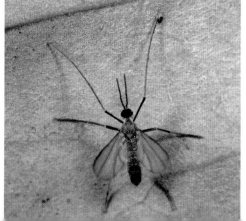

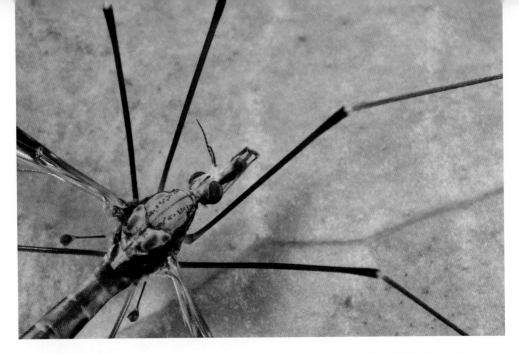

Crane flies have particularly prominent halteres – the second pair of wings reduced to these club-like balancing organs which define all flies.

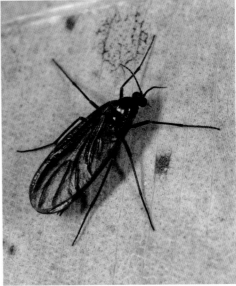

The **Mycetophilidae** is a family of about 4,000 species, known as **fungus gnats**. Most are associated with fungi as larvae. The species on the left is swarming around a fresh mushroom to mate and lay eggs. On the right is a species of a smaller related family, the **Sciaridae**, known for good reason as **black fungus gnats**. Both from Australia and about 0.5cm (0.2in).

Not your ordinary fly. While the majority of **midges** in the family **Cecidomyiidae** are small and dull, there are fabulous exceptions to the rule. Here a species of (probably) *Asphondylia*, resting on a *Heliconia* in the highland rainforest of Costa Rica. 1.8cm (0.7in) long.

Now for the **midges** that do bite. The **Ceratopogonidae** has hundreds of minute species, especially in the genus *Culicoides*, with mythological status. In North America they are called no-see-ums, punkies or sand flies. In Europe they are known as biting midges. At less than 0.2cm (0.06in), they are nearly invisible, but their bite causes severe itching because they rasp away skin and then spit saliva into the cut before drinking the blood. Only wet mud is needed to breed astronomical numbers, and the adults attack in swarms. A volunteer study in Scotland clocked a record biting rate of 635 bites on one arm in one minute! Shown here are a Scottish monster on the left and a greedy Australian species carrying three times its weight in blood on the right.

Black flies, family **Simulidae**, are another biting fly group. A few are so extreme that they can kill an animal by blood loss alone. Like the no-see-ums (top of page), they rasp away skin and spit anticoagulant saliva, causing severe itching. Left is an *Austrosimulium* species from New Zealand, and right is *Simulium bicoloratum* from Ecuador. Both about 0.3cm (0.1in).

The two tiny flies on this page are in the **Psychodidae** family. Below is a species from a group aptly known as moth flies, or filth flies, which are common in bathrooms and kitchens, where they breed on slime in the plumbing but do not bite. The species above .is from a group known as sand flies, *Lutzomyia* sp., which do suck blood and can spread diseases including leishmaniasis. Both about 0.3cm (0.1in).

Members of the suborder **Brachycera** are characterised by short antennae and stouter bodies. These are the more modern **flies**, in more than 120 families, and include all the 'typical' forms such as house flies. Fewer families have aquatic larvae, but most possible terrestrial lifestyles are employed. The evolutionary sequence starts with the horse flies and related species, the Tabanidae.

The family **Tabanidae,** with 4,400 species, contains many nuisance flies, known as **horse flies**, **deer flies**, **clegs** and **march flies**. Their impressive rasping mouthparts are mounted on a long proboscis, and most species painfully bite animals for blood. The larvae are voracious hunters of other invertebrates, living in hidden moist habitats. Adults are medium to large, up to 3cm (1.2in). Above and below are species from Australia, including. *Scaptia* sp. above, both 1.6cm (0.6in).

Horse flies are very annoying biters, but some are also beautiful insects, such this rainforest dweller from Ecuador, 1.5cm (0.6in).

Most species of '**horse flies**' have no access to horses, but bite a variety of mammals, including humans. The huge compound eyes can have more than 5,000 individual facets to hone in on prey. This large species is from high-altitude forests in New Guinea. 2cm (0.8in).

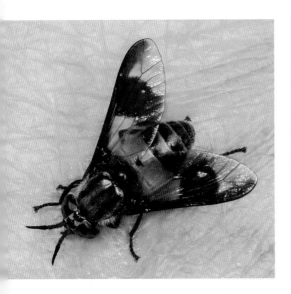

Deer flies are another part of the **Tabanidae** family, and many are associated with biting deer. However they also love to bite humans in America and Europe. They are smaller than most horse flies, with markings on the wings and spectacular eyes. *Chrysops viduatus* from Europe, 1.4cm (0.5in).

Below right: A **deer fly** from Malyasia, again showing the amazing patterns across the compound eyes. 1cm (0.4in).

The spectacular eyes on this **horse fly** from Poland are not what they seem. They are made up of several thousand separate facet eyes, and the red 'pupil' is just pigment across some of the facets.

The colourful eyes of some flies are just external reflective pigment, theoretically not affecting their internal colour perception. **March fly**, *Lissimas australis*.

One of the prettiest flies in the Australian alpine country is a pollinator, but also a nasty biter. **March fly**, *Scaptia auriflua*, 1.2cm (0.5in).

Snipe flies are a small family, the **Rhagionidae**, related to the horse flies. Few species are blood drinkers, the common European species on the right, *Chrysopilus* sp., apparently does not feed as an adult at all. On the left is a larger species from Costa Rica, probably a biter, with the characteristic horse fly-type proboscis. 1cm (0.4in) and 1.5cm (0.6in).

Soldier flies, family **Stratiomyidae**, are distinctive, bright flies with about 3,000 species. Their bodies are somewhat flattened, and eyes are often very large and multicoloured. The larvae are mainly scavengers and algal eaters in water and damp places, although some are predators and have a distinctive armoured body different to the more usual maggot-like larvae of other flies. Size varies from a tiny 0.3cm (0.1in) to around 2cm (0.8in).

Soldier fly laying eggs above a damp piece of forest in Madagascar. The larvae will drop down when hatched. 1.2cm (0.5in).

A variety of mostly unidentified **soldier fly** species, **Striatiomyidae**, from various tropical locations.

From Indonesia, 1cm (0.4in).

From Malaysia, 1.2cm (0.5in).

From New Guinea, 0.9cm (0.4in).

From Thailand, 0.8cm (0.3in).

From Thailand, 0.8cm (0.3in).

Syndipnomyia auricincta is one of many species of **soldier flies** which mimic wasps. It gives them a small amount of protection, unless too many mimics are in an area and the predators forget to be careful. From Australia, 1.2cm (0.5in).

Oplodontha viridula is a common European **soldier fly**, known as the 'green colonel'. Note the eye patterns. 1cm (0.4in).

Members of the family **Asilidae** are known as **robber flies** and they are great hunters. Their long muscular bodies, big eyes and spiked legs give them a comparable hunting advantage to the dragonflies. There are more than 7,000 species of these handsome flies, ranging in size from 0.3cm (0.1in) to monsters over 6cm (2.5in). They catch insect prey on the wing, impaling it on their beak-like mouthparts, and aided by spines on the legs. The larvae are also predators, often attacking scarab beetle larvae in the soil.

The huge wrap-around eyes of **robber flies** provide early warning of approaching prey. Although they can carry their own weight impaled on the 'beak', this Malaysian species is demonstrating this method with a small snack of a *Drosophila* fly victim. 2cm (0.8in).

A beautiful monster, this **robber fly** from the genus *Chrysopogon* is from arid Western Australia where it hunts flying insects. 2.8cm (1.1in) in length.

Opposite: It may not be the most eye-catching **robber fly**, but it deserved the full page to convey its surprising scale. This 6cm (2.4in) long hunter was easily flying with a full-sized African locust impaled on its beak. From the arid Madagascar interior.

An eye-catching **robber fly** from Australia, 1.4cm (0.5in).

Choerades gilva is part of a genus of hairy **robber flies** found mainly in Europe. 1.8cm (0.7in).

This desert **robber fly** is from New Mexico, USA, where it had caught a horse fly for a meal. Robber flies can be beneficial insects. 2.2cm (0.9in)

Not all **robber flies** are stout-looking hunters. This delicate long *Leptogaster* sp. is found in South-East Asia and Australia. 2cm (0.8in).

The **robber fly** *Blepharotes coriarius* is one of the biggest insects in Australia. This handsome bearded beast is 4.2cm (1.7in) long. They deposit eggs into cracks in dead eucalyptus trees, or in cracks in the soil, where their predacious larvae seek out mainly beetle larvae.

Robber flies are fearsome enough, but many species have also adapted to mimic wasps and bees, which some predators learn to be scared of. Top left is *Chrysopogon* sp., 1.5cm (0.6in), from Australia. Top right is a mimic of giant carpenter bees from Ghana, 3cm (1.2in).

Casually hanging by one claw, this female is eating her prospective mate. **Robber fly** males often present a female with an insect to feed on during mating. This male may have omitted the gift, and become one himself. 2.6cm (1in) long each.

Two **mydid flies** from Western Australia, the key location for the family **Mydidae.** Right a mating pair of *Diochlistus mitis*, 2cm (0.8in), and left *Miltinus stenogaster*, 2.5cm (1in).

Two more **mydid flies** from their Western Australian semiarid habitat. Left is a mating pair of *Miltinus maculipennis*, mimicking typical wasps, 1.8cm (0.7in). Right is a possibly undescribed species of *Miltinus*, 2cm (0.8in).

The **Apioceridae** is a small family of only 140 species, with half of them occurring in arid parts of Australia. Adults frequent flowers, with marvellous hummingbird-like hover-feeding. Larvae are predators in the soil. Both species belong to the single family genus of *Apiocera* and are about 1.8cm (0.6in) long.

Stiletto flies, family **Therevidae**, are about 1,000 species of slender flies. Little is known about their adult habits, but the long and thin larvae are big time hunters of other invertebrates in the soil. Arid Australia is again the key location for this group.

Stiletto fly adults use sugars from flowers for their food, but are rarely seen doing so. *Medomega averyi* from Western Australia is caught in the act. 1.2cm (0.5in).

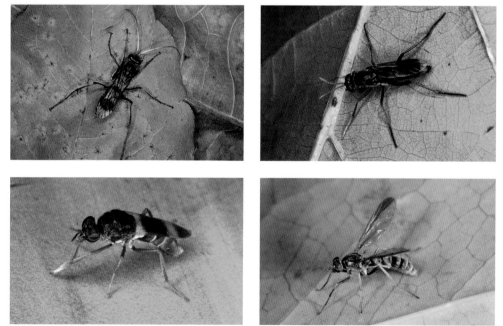

Stiletto flies have a number of wasp-mimic species, sometimes just in general, such as *Taenogera luteola* (above left) and *Pipinnipons fascipennis* (above right). Sometimes they impersonate a specific species. Here *Agapophytus omaticornis* (top right) is doing a good job of looking like the fearsome spider wasp, *Auplopus* sp., Pompilidae (top left). All from Australia and about 1.5cm (0.6in).

The **bee flies**, family **Bombyliidae**, is a large family of often fluffy flies, sometimes generally mimicking wasps and bees, and sharing their habitat of hovering around flowers. They lay huge numbers of tiny eggs in places where hapless insects will pick up the tiny hatchling larvae, which are parasitic. Bees, wasps, antlions and grasshoppers, especially as pupae and larvae, are the major hosts. More than 5,000 species of bee flies are major pollinators, more so than bees in some arid habitats. They are seen hovering around flowers, sometimes with very long pointed 'beaks' to inject deep into the flowers.

Members of one group of **bee flies** do not have the fat fluffy body but this skinny short haired form, with extra-sharp stiletto mouthparts to stab deep into flowers. 0.8cm (0.3in).

Brown and grey hues dominate **bee fly** coats, but a few stand out with more garish palettes. From Indonesia, 1.6cm (0.7in).

Bee flies often lay their eggs on the ground and can be seen hovering and stopping. Some stay on the ground for a long time and pick up sand into a special body chamber where it coats the moist eggs. As the female deposits them onto similar sand, the eggs are virtually invisibile. *Villa halteralis* from Europe, 1.5cm (0.6in) body.

Some of the largest **bee flies** in Australia are in the genus *Comptosia*. Shown here is a very heavy species, *Comptosia magna*, with a 5cm (2in) wingspan, which seems hardly enough to lift its fat egg-laden body off the ground, where it lays a single egg per landing. Bee flies often have the habit of landing with their wings fully extended.

In their quest for sugars **bee flies** can get become covered in pollen, as this South African *Anastoechus* sp. shows. It makes them important pollinators in the arid habitats they frequent. 1cm (0.4in).

A black *Bombylius* sp. of **bee fly** from Australia shows the almost fighter-plane symmetry and large thorax full of flight muscle, which are typical of these strong fliers. 1.5cm (0.6in) body.

A good pose to show off the straight 'beak' of **bee flies**, used to dig deep for sugars in flowers. *Disodesma* sp. from Western Australia, 1.5cm (0.6in) body.

The fluffy body of a very common **bee fly** from the veld flower belt of South Africa. With thousands of species of flowers in this world botanical hot-spot, bee flies and many other pollinator insect groups have a huge food resource. *Anastoechus* sp., 0.8cm (0.3in).

One of the hordes of **bee flies** which pollinate the extreme flower diversity of Namaqualand, South Africa. Here honey bees are not needed as hundreds of species of native bees and flies are adapted for the task.

The taxonomy of these tiny mating flies is in flux. Their common name is 'micro bee flies', as most are 0.3cm (0.1in) or less. Some authors ascribe a family for them, the **Mythicomyiidae**, while others lump them as a subfamily of the larger bee flies, the **Bombyliidae**. They have similar habits, so are often found near flowers. From Western Australia.

Dance flies, family **Empididae**, are known for their mating swarms, where each male delivers a captured insect as a nuptial gift to a female. This 1cm (0.4in) Costa Rican species is eating a vinegar fly.

This tiny 0.3cm (0.1in) fly is known as a **smoke fly**. It is attracted to dying camp fire smoke, and other flies seeking food there. *Microsania* sp., **Platypezidae**.

The family **Dolichopodidae** contains about 7,000 species of the '**long-legged flies**'. They are small, usually under 1cm (0.4in), but often strikingly shiny metallic, with slender bodies and a high stance at rest. Most are predators of smaller insects, and their larvae are also predacious in the soil and under bark.

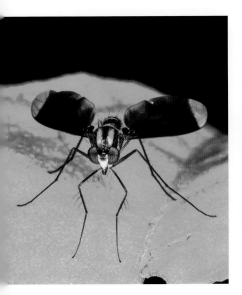

One of the larger examples of **long-legged flies**, a 1.2cm (0.5in) species from the highland forests of New Guinea.

The **long-legged flies** are most commonly green, which makes this yellow hunter, an *Amblypsilopus* sp. from Australia, quite unique. 0.8cm (0.3in

Long-legged fly from Borneo, 1cm (0.4in).

Two **long-legged flies** from Borneo. The side-on view shows how they got their name. These flies are territorial and constantly re-land on the same horizontal leaves from which to pounce on passing insects. Both about 1cm (0.4in) long.

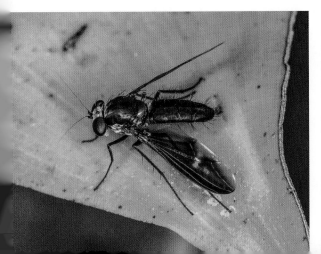

The **Syrphidae** is a large family of more than 6,000 species, known as the **hover flies**. In temperate climates spring time is abuzz with these hovering, flower-visiting flies. Many have general wasp-mimicry patterns, and some are very specific mimics of wasps and bees. Their larvae are hunters of soft insects such as aphids, and therefore a gardener's friend, while others parasitise native bees and more.

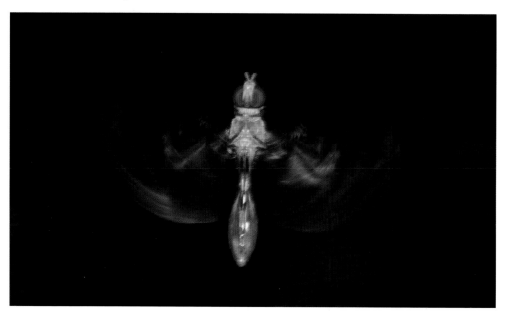

Hover fly from Ghana doing what it does so well, hovering on the spot. *Baccha* sp., 1.2cm (0.5in) body.

The **hover fly** genus *Volucella* has many large handsome species, with varying degrees of wasp and bee mimicry. *Volucella zonaria* is a general hornet mimic from Europe, 1.6cm (0.6in).

Look carefully at flower-visiting insects. Many wasps, bees and bumble bees are actually convincing imposter **hover flies**.

Criorhina floccosa, bumble bee mimic. Europe, 1.5cm (0.6in).

Chrysoroxum cautum, hornet mimic. Europe, 1.5cm (0.6in).

Eristalis intricarius, bumble bee mimic. Europe, 1.6cm (0.6in).

Volucella bombylans, bumble bee mimic. Europe, 1.6cm (0.6in).

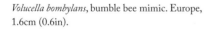

Ceriana ornata, wasp mimic. Australia, 1.2cm (0.5in).

Ceriana conopsoides, wasp mimic. Europe, 1.4cm (0.6in).

One of the best insect mimics in the business. This is not a *Vespula* sp. hornet, but a harmless **hover fly** pollen eater, *Temnostoma vespiforme*. The Latin name says it all. Europe, 2cm (0.8in).

This **hover fly** is a very convincing honey bee mimic from Thailand, 1.2cm (0.5in).

A rare black species of **hover fly** from Western Australia, *Orthopsopa grisa*. Its larvae are predators of ants. 1.2cm (0.5in).

Austalis pulchella, a **hover fly** with metallic sheen from Australia, 1.4cm (0.6in).

Large-headed **hover fly** tasting a leaf with traces of fruit juice dripped from the forest canopy, Costa Rica, 1.2cm (0.5in).

Hover fly species of the genus *Eristalinus*, from tropical Asia and Australia, always amaze with the patterning of their huge eyes. 1.5cm (0.7in).

The **hover fly** *Austalis conjuncta*, from tropical Australia, is commonly noticed at flowers thanks to its extreme red eyes, 1.4cm (0.6in).

Deep metallic green **hover fly** from Costa Rica, *Omidia* sp., which mimics orchid bees in its general appearance and flight mode. 1.5cm (0.7in).

Blossoms can be hard to come by in the rainforest. These two **hover flies** are pollinating a tree in the Australian World Heritage forests, and doing what they do best, hovering. 1cm (0.4in) long.

Members of the **Ephydridae** family are called **shore flies**, but they are not restricted to this habitat. Like many insects, they can walk on water with the aid of hygroscopic hairs on their feet. Left, an Australian species devouring a midge on the surface, and right, a species from New Guinea with raptorial forelegs, like those of mantids, for catching prey. Both about 0.8cm (0.3in).

Members of the **Sciomyzidae** family are known as **snail killers**, as they actively hunt and parasitise many types of freshwater molluscs. Adults can be found lurking on water plants. 1.4cm (0.5in).

Although not as famous as wasps for this task, many fly groups are parasitic, and beneficial to our crops. *Cryptochaetum iceryae* is a 0.2cm (0.08in) fly, which has been exported from Australia around the world to control the crop pest, cottony cushion scale. Here it is laying eggs into the host. **Cryptochaetidae.**

The **Conopidae** is a parasitic family of strongly patterned flies, often mimicking the wasps and bees that they parasitise. Even the waist of the wasp is emulated. Most of the 800 species are seen at flowers where they can jump arriving wasps and bees, and using a special 'can opener' like device at their rear, prise open a gap in the abdomen of the host and insert a single egg. Hosts are eventually killed by the feeding larva.

Two good wasp mimics. Left, the Australian *Physocephala australiana* looking like many mud wasps, and right, *Conops vesicularis* from Europe, generally mimicking hornet types.

A part of this family are good pollinators, and even have very long stiletto-like mouthparts to delve deep into many flower species, as *Sicus ferrugineus* shows on the left. On the right is a general wasp mimic, *Conops vesicularis*. Both from Europe, and about 1.5cm (0.6in).

A handsome general wasp mimic from Australia, *Australoconops* sp. (left), and a general wasp mimic, *Physocephala biguttata* from the Canary Islands (right). Both about 1.5cm (0.6in).

A wasp-mimic fly from Sri Lanka, filling up on sugars, the main fuel for insect flight muscles. Wingspan 2.5cm (1in).

Many flies have the words 'fruit fly' in their name, but the **Tephritidae** family are the 'true' **fruit flies**. With more than 4,600 species, it has some famous pest species feared by orchard and other farmers. However, as always these are just a minute subset of species, and most members of the family do not trouble humans. Most larvae develop inside plant tissue, especially fruit, from eggs stabbed into the plant. Shown here are: the famous Mediterranean fruit fly, *Ceratitis capitata* (above left); the papaya fruit fly, *Bactrocera papyae* (above right); and the mango fruit fly, *Bactrocera frauenfeldi* (below). Most of the pest species are in this very large genus, and around 1cm (0.4in) long.

Most **fly larvae** are called maggots, and here are the maggots of a typical fruit fly, *Bactrocera papayae*, eating a capsicum from the inside out.

The green bamboo **fruit fly** lives in South-East Asian forests, and the maggots feed on the soft flesh of the growing tips. 0.8cm (0.3in).

Austronevra bimaculata is a **fruit fly** native to Australian rainforest. Here the male is performing a mating dance, which is common among fruit flies. Note the long egg-laying organ in the female below, used to pierce fruit.

An unidentified large **fruit fly** species from the rainforests of New Guinea, 1.2cm (0.5in)

A tiny **fruit fly**, 0.4cm (0.16in) long, from Indonesia.

A giant **fruit fly** from the rainforest of Ghana, with a wingspan of 3.5cm (1.4in). Family **Tephritidae.**

Another fly family with many species called fruit flies is the **Platystomatidae**. Their more normal name is **broad-mouthed flies,** and it is easy to see why on this species staring at the camera in New Guinea. 2cm (0.8in) wingspan.

Among the **Platystomatidae**, are antler flies in the genus *Achias*. Above is a female with no eye stalks, and below a species from New Guinea with one of the widest spreads. Both body length about 1.5cm (0.6in) long.

Another name for the broadmouth flies, **Platystomatidae**, is signal flies. Like other fruit flies, they use their elaborate wings to signal during courtship. Species from New Guinea, 1.5cm (0.6in).

A **signal fly**, *Euthyplatystoma* sp., **Platystomatidae**, from Malaysia. Apart from the outrageous eyes, this family of around 1,200 species is often adorned with very intricate wings, which are used by the males in elaborate courtship dances. 1.5cm (0.6in).

The **Richardiidae** is a small **fruit fly** family mainly from South America, where members specialise more on bird and animal droppings than fruit. 1cm (0.4in).

An unidentified stalk-eyed **signal fly** species from New Guinea. This hammerhead is not just to impress the normal-headed females, but for fighting with other males. Body 1.5cm (0.6in) long.

Not all **signal flies** have the very robust and patterned wing appearance. This species from Indonesia breaks from the norm with a small head and a wasp-like waist. 1.5cm (0.6in).

The weirdest fly of all? Barely looking like a fly, this *Asyntona* sp. is a **broad-mouthed fly**, family **Platystomatidae**, possibly mimicking a beetle. From Australia, 0.5cm (0.2in).

The family **Drosophilidae** is best known for the **vinegar flies**, *Drosophila melanogaster*, the tiny flies that hover above fruit bowls in houses all over the world (below left). This is also the insect whose genome has been studied more than any other. However this family has more than 4,000 species not found in houses. Their main food is yeasts, fungi and plants, while some are predators. Shown here is a species of *Drosophila* on a mushroom in Borneo (bottom left) and *Leucophenga scutellata*, a gaudy species from Australia (below right). All 0.4–0.5cm (0.16–0.20in).

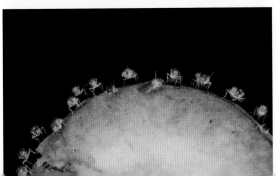

The **Lauxaniidae** is a fly family that defies description, other than that most of the 2,000 small-sized species are surprisingly intricate and beautiful in close up. The lifecycle of many species is unknown, but some larvae feed on decaying leaves, sometimes living between the layers.

Top left: *Sapromyza* sp. from Australia, 0.8cm (0.3in).

Top right: *Cephaloconus* sp. from Australia, 0.8cm (0.3in).

Middle left: Metallic fly from New Guinea, 0.4cm (0.16in).

Middle right: *Cerastocara* sp. from Australia and New Guinea, 0.8cm (0.3in).

Left: Species from Australia, 1cm (0.4in).

The two flies above are as odd as they come. The 120 species of the family **Celyphidae** are known as **beetle flies** for obvious reasons. A part of the thorax is so enlarged that it forms a shield over the single pair of wings, much like real beetles with fused wing cases. They are mainly found in Asia and the larvae feed on decaying plants, including grasses. Up to 0.5cm (0.2in).

This handsome fly is a rare species in the **tangle-winged fly** family **Nemestrinidae**. Some species have very a long proboscis and hover hummingbird style pollinating deep flowers, but this one, *Nycterimorpha* sp., called the 'bat fly' has unknown habits in Australia. Wingspan 2.2cm (0.9in).

Antler flies are about 200 distinctive species in Asia and Africa, in the family **Diopsidae**. Their eyes and antennae are on often very long and slender antlers. The unrelated **stalk-eyed flies**, family **Platystomatidae**, have their antennae still in the middle of the more robust hammerhead like stalk. The antlers are found on both sexes, although the male ones are longer and are used in combat over females, who tend to prefer the males with the longest antlers.

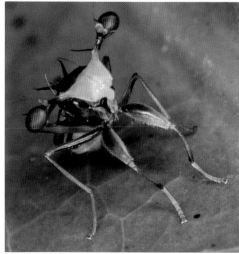

One of the most extreme species of **antler fly**, from Borneo, with the antlers longer than the body. 1.2cm (0.5in).

African **antler flies** often have shorter antlers and bigger eyes, and lots of attitude. This one is from Ghana, 1cm (0.4in).

A Bornean **antler fly**. Note the four spines on its back. One theory is that they help to guide the bodies into the right stance for mating (see right).

A mating pair of **antler flies** showing that sometimes the sexes have similar sized antlers. From Thailand, 0.8cm (0.3in).

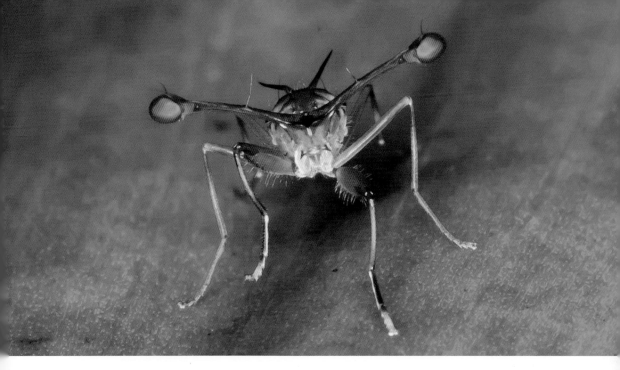

An African species of **antler fly** from Ghana. Look closely at the middle of the head which has an interesting accidental illusion of an angry bearded face. 1.2cm (0.5in).

A mating pair of **antler flies** shows that the females also possess antlers, although often smaller. From Malaysia, 1cm (0.4in).

The **stilt-legged flies**, about 700 species in the family **Micropezidae**, have very long middle and hind legs, and are prominently seen on horizontal leaves especially in the tropics. They have complex courtship dances and food exchange that can look like kissing, and spend much time in the mating position. Known larvae are found in decaying plants, and adults are attracted to rotting fruit and animal droppings.

Nestima sp. from Indonesia, 1.5cm (0.6in) long.

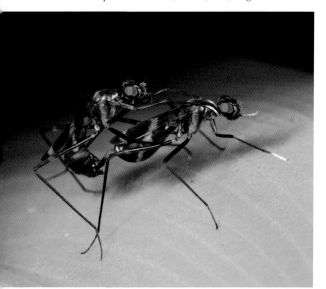

A mating pair of *Mimegralla* sp. from Australia, 1.5cm (0.6in) long.

Stilt-legged fly from Borneo, 1.5cm (0.6in) long.

Stilt-legged fly from Indonesia, 1.4cm (0.6in) long.

The 'Calyptrate' flies are a group of families that most personify what people mean when they use the word fly. The house flies, bush flies, blow flies and flesh flies. We think of them as pests because of the few species that are always close to humans, and feed on 'waste'. Most of these flies are saprophages, meaning they recycle decaying matter, from faeces to dead bodies – a job absolutely necessary to keep ecosystems working. And even among the house fly family, **Muscidae**, many are hunters of other insects, including other pest species.

The family **Muscidae** contains several pesky species. In parts of Australia it is impossible to enjoy the outdoors in summer because *Musca vetusticima,* the **bush fly** (above left), lands everywhere it can get a bit of moisture. The worldwide **house fly**, *Musca domestica* (above right), brings all manner of diseases from rubbish and faeces into our homes. Both about 6–8cm (0.2–0.3in) long. Spare a thought for the poor cow attacked by the ferocious **horn flies**, *Haematobia irritans* (below left), originally from Eurasia but now spread around the world. Below right is a close-up of the fabulous biting mouthparts of the **stable fly**, *Stomoxys* sp., which has both sexes biting for a blood meal. Their larvae live in urine-soaked hay, hence stables are their favourite habitat, where they quite like to bite humans too.

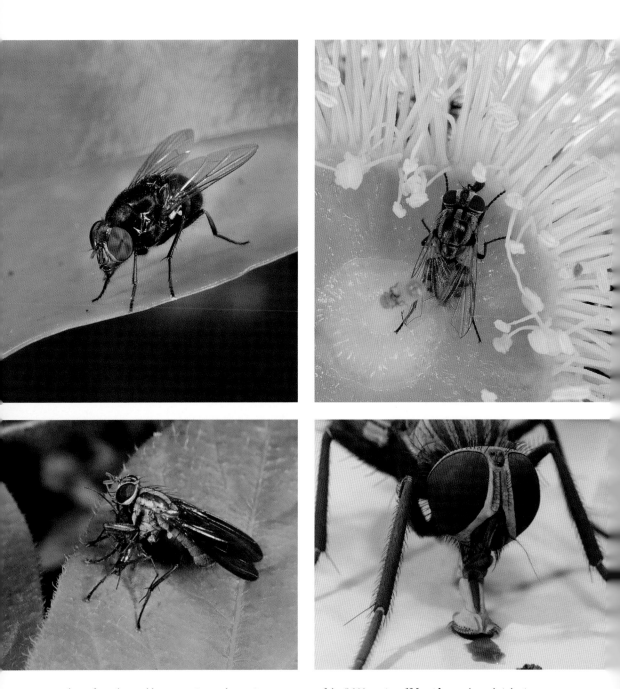

Apart from the notable pest species on the previous page, most of the 5,200 species of **Muscidae** go about their business out of our lives. Top left is *Neomyia* sp., living in rainforests and recycling wastes there. Top right shows a **muscoid fly** pollinating flowers, and bottom left is one of the many species which hunt other insects. Bottom right shows the 'mop-like' mouthparts of this group, which easily transmit slops and bacteria between rubbish and faeces and food in homes.

The **blow flies**, family **Calliphoridae**, are a group of about 1,600 species of flies known for their meat-eating habits on carrion and damaged flesh. More species are parasites of other invertebrates, but the concept of an animal, like a sheep, being 'fly-blown' is hard to ignore. This genus is found worldwide and some species are studied by forensic experts to determine the 'age' of dead bodies. Left is a species of the genus *Chrysomya*, which includes the notorious **screwworm flies** that attack the living flesh of mammals including humans. Both 1–1.2cm (0.4–0.5in).

Neocalliphora albifrontalis, a **blow fly** from Western Australia, sitting on the feathers of a dead emu, where its maggot larvae will develop. 1.2cm (0.4in).

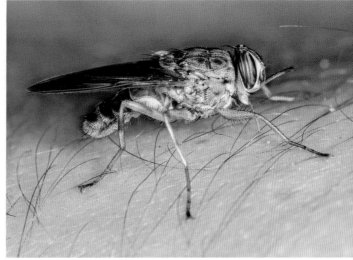

This is the legendary African **tsetse fly**, *Glossina* sp., in its own family, the **Glossinidae**. This viscious biter of large animals, from rhinos to cattle to humans, carries the Trypanosoma parasite which causes sleeping sickness. Both sexes bite, but females then use the blood to rear one larvae at a time to maturity inside their bodies. Unlike many biting insects, once you have been sighted, it will chase you.

Many species of **blow flies** are metallic blues and greens – the origin of the names bluebottle and greenbottle. This one is from Malaysia. 1cm (0.4in).

The family **Sarcophagidae**, known as **flesh flies**, have similar habits to the blow flies. Most of them are distinctive by having grey bodies with three dark stripes along the thorax. Shown here is a European species of *Sarcophaga*. 1.1cm (0.4in).

Stomorhina xanthogaster from Australia is an example of a **blow fly** which does not have anything to do with meat. It is one of the many parasitic species, with larvae developing among ants and termites. 1cm (0.4in).

An Australian **flesh fly**, *Sarcophaga* sp., attracted to the flesh of a stink flower, whose perfume is more rotting meat than Elizabeth Arden. 1.2cm (0.5in).

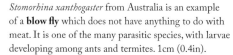

The last major family of flies in the evolutionary sequence followed here is the **tachinid flies**, family **Tachinidae**. This huge group of more than 10,000 species are all parasites of mainly plant-eating insects. Their role in the environment is massive, and some of their prey are pests of our food crops. Eggs are normally laid on the prey, especially caterpillars, beetle larvae and bugs, and the maggots burrow in to develop inside, killing the prey before it develops into an adult. There is a great variety of shapes and sizes, but most have prominent bristles at the rear and other parts.

This gory mess is a pupa (chrysalis) of a killed birdwing butterfly, opened to reveal a maggot and pupa of the parasitic **tachinid fly**, *Blepharipa* sp., in Australia.

Tachinid flies have many shapes, and a classic one is a broad fat body, often with bright patterns, as in this species from Indonesia. Body 2.5cm (1in) long.

A tense scene. The very hairy caterpillars are those of processionary moths, covered in hairs which do not just irritate but are poison-tipped. The parasitic **tachinid fly**, with egg-laying ovipositor extended, is waiting for a chance to jump in between the hairs and deposit eggs onto the skin of the caterpillars. Fly 1cm (0.4in).

Tachinid flies have many shapes, and a classic one is a broad fat body, often with bright patterns, as in this species from Indonesia. Body 2.5cm (1in) long.

Two **tachinid fly** species from Europe. Left: *Phasia aurigera*, which parasitises pantatomid shield bugs.
Right: *Gymnocheta viridis*, which comes in a variety of shiny metallic forms. Both about 1.2cm (0.4in) long.

Prodiaphania sp., an Australian **tachinid fly** parasite of scarab beetle larvae, must be one of the broadest flies, appearing almost square from above. 1.6cm (0.6in) long.

The other extreme of body styles among **tachinid flies** is this skinny round-bodied *Cylindromyia* sp. When not parasitising stink bugs it fuels up on sugars from flowers. 1.2cm (0.5in) long.

An unidentified **tachinid fly** from the islands of New Guinea does not skimp on brightness. 1.2cm (0.5in) long.

Several genera of **Tachinidae** possess extra bristles, to the point that the name '**hedgehog flies**' applies. They are largely tropical, and this species is from the Costa Rican highlands. 2.2cm (0.9in) long.

This handsome **hedgehog fly** is from high-altitude forests in Costa Rica – a habitat favoured by many species. It is probably one of the many species of the genus *Epalpus*, which are parasites of caterpillars. 2.2cm (0.9in) long.

The last family of flies here is the bizarre **Hippoboscidae**. It has 800 species of ectoparasitic flies, about 600 of which live only on bats, and the rest on birds. Left: A species of **bat fly** the genus *Cyclopodia*, from the subfamily **Nycteribiinae**, on the fur of a flying-fox in Australia. It is so perfectly adapted for this life in fur that it has lost many normal fly features and is mostly blind. Right: This **streblid fly** on bat fur is member of the other subfamily, the **Streblinae**, most of which have wings, although some lose them after finding a host. Their coevolution with their hosts often results in a bat fly species being linked to only one bat species host.

CADDISFLIES

Order TRICHOPTERA

14,000 species in 45 families

Caddisflies are well known to fisher people and anyone who frequents freshwater habitats. It is a small order, with only about 14,000 species in the world, but with very wide distribution. They are closely related to moths and butterflies. The main differences are lack of scales on the wings, lack of the curled moth proboscis, and the caddisfly antennae are very long and not hairy. The aquatic larvae of caddisflies, with elaborate cases, traps and hides, are not typical of moths either, although there are aquatic moth caterpillars in one family.

The size range is surprising. A typical species seen dancing above water may be about 1.5–2cm (0.6–0.8in) in length, but they vary from a minute 0.15cm (0.06in) to a large 3.5cm (1.5in) long.

Adults swarm in the summer, dancing above the water in mating rituals. Females make a large number of tiny eggs which are laid in sticky strings, much like frogs' eggs. These hatch into very active larvae, most of which build cases out of leaves, sand or sticks, with particular designs being diagnostic to the family or genus. Detritus and other plant matter are the main food, but the larvae which do not build cases make nets in flowing water which capture small prey for their carnivorous habit. The majority of larvae have gill tufts for breathing, but some can respire directly through their 'skin'.

Adults eat little or not at all, living short lives spent in creating the next generation. Caddisflies can inhabit clean waterways in huge numbers, and so are an important part of the food chain. Young fish especially rely on their presence for food, and the adults are prey for frogs, birds, bats and fish when they fall on to the surface.

Typical **caddisflies** are like this Australian species, *Triplectides* sp., family **Leptoceridae**. Brownish with slightly hairy wings and very long thin antennae. It lives in fast streams in the tropics. 2.5cm (1in) long.

They are sometimes confused with the other main aquatic orders. The mayflies differ by having the three cerci, or long tails, even as larvae. Adults tend to look fairly drab, although gaudy species exist in the tropics. Their forewings are covered by hairs, and this can make adults at rest look a bit like elongated moths.

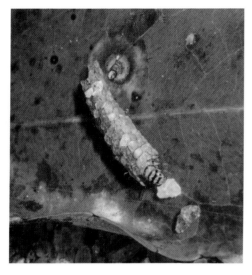

Caddisfly larval cases made of sand are common among several families, especially in very clear streams with little debris. Apart from being good general camouflage, some species actually create snail shell-shaped ones for a different disguise.

Among the more ingenious shelters are those contructed by larvae of the **Philopotamidae** family of **caddisflies**. They make fine nets, of silk, algae and sand, placed open into the current, either between stones or under stones. The larva sits safe at the end, capturing algal particles and protozoa as they are flushed in.

Leaves are a common material for **caddisfly larvae cases**. They require little silk to glue together, and in streams with large leaf loads are excellent camouflage. This one belongs to *Triplectides* sp., family **Leptoceridae**. See adult on opposite page.

Reeds and sticks are the other common material for **caddisfly larval cases**. They are more intricate to weave and glue together, but again good camouflage among stream debris. All the cases here are added to as the larva grows.

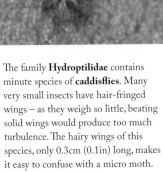

The family **Hydroptilidae** contains minute species of **caddisflies**. Many very small insects have hair-fringed wings – as they weigh so little, beating solid wings would produce too much turbulence. The hairy wings of this species, only 0.3cm (0.1in) long, makes it easy to confuse with a micro moth.

The genus *Stenopsychodes*, in the family **Stenopsychidae**, contains many of the more ornate **caddisflies** in the Asian and Australian region. 1.8cm (0.7in) long.

Below: *Anisocentropus bicoloratus*, in the **Calamoceratidae** family, is one of Australia's most eye-catching **caddisflies**. The larvae have leaf cases in slow streams. 1.2cm (0.5in) long.

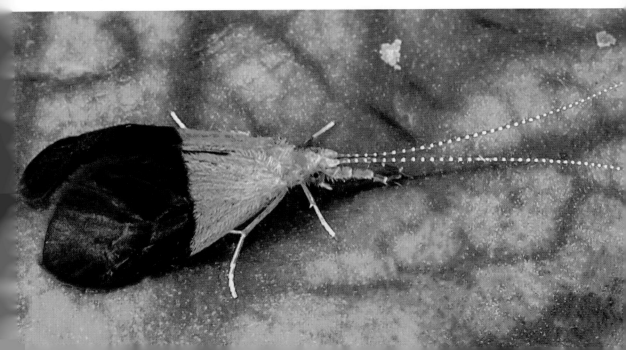

Aethaloptera sexpunctata is in the **Hydropsychidae** family, which contains more than 900 species of **caddisflies**, some of which have shiny, hairless wings, making them easy to confuse with lecewings. 2cm (0.8in) long.

A house of stone made by the larvae of a **caddisfly** in the **Glossosomatidae** family.

A house made of reeds by an American species of **caddisfly** of the family **Limnephilidae**.

Left: The sedge fly is a **caddisfly** very well known to anglers in America, where it is imitated as a lure in stream fishing. *Pycnopsyche* sp., **Limnephilidae.**

Below: One of the many egg-laying strategies among **caddisflies**. This gelatinous mass will soon release the hatchlings into the stream below.

MOTHS and BUTTERFLIES

Order LEPIDOPTERA

175,000 species in 126 families

Moths and butterflies need little introduction in terms of identification. The general body form of this large order is two pairs of wings, covered in overlapping scales rather than hair. The name is derived from the Greek *lepis*, meaning scale, and *pteron*, meaning wing. Some of the mouthparts are fused into a coiled tube called a proboscis, although the palps are retained and can be quite large and tusk-like. Some moths, like the hawk moths, which pollinate deep-throated flowers, can have a proboscis several times longer than the body. The antennae are shorter than the body in most species, and can often be fan-like or 'fluffy' in the males. Moths are very closely related to caddisflies (previous chapter), and the most primitive moth families still have the same chewing mouthparts as caddisflies.

The **Micropterigidae** family, illustrated here by a member of the genus *Sabatinca*, has the oldest lineage of **moths**, dating back at least 120 million years. They have chewing mouthparts rather than a proboscis. 0.6cm (0.2in) long.

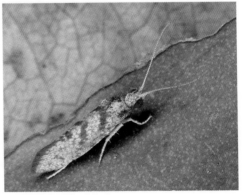

The other very primitive **moth** family, with chewing mouthparts, is the **Agathiphagidae**. Their structure is similar to caddisflies, with which they share ancestry. This tiny Australian species, found in kauri pine cones, is one of only two known in the world. 0.4cm (0.15in) long.

Ghost moths, family **Hepialidae,** are medium to very large, often spectacularly marked, but rarely seen. This is because they are rarely attracted to light, and only live for one day as adults! During this time they must find a mate and lay eggs. Larvae burrow for tree roots or in tree stems. On the left is *Aenetus mirabilis*, 6cm (2.4in), and on the right is *Abantides hyalinatus*, 5cm (2in), both from Australia.

Size variation among so many species is extreme. From tiny moths among the **Nepticulidae** family, with a wingspan of only 0.3cm (0.1in), to monsters among the Atlas and Emperor moths with wingspans up to 30cm (12in).

In general all species have plant-eating caterpillars. The larger ones tend to feed in the open, and sometimes use camouflage, warning patterns or irritant- and poison-barbed bristles, to protect themselves from general predators. Specialised predators/parasites, such as the paper wasps, hornets and other wasps, are not deterred by these methods. Small caterpillar species tend to live a more hidden life, with leaf mining a major strategy. Hiding between the layers of a leaf is relatively safe. Just a few species have developed hunting habits and feed on insects such as aphids.

The pupa stage in moths tends to be a brown lozenge shape which is hidden shallowly in the ground, under bark or in leaf litter. Some butterfly pupa, known as a chrysalis, are often hung from branches, although only the potentially poisonous ones are very exposed, and tend to have bright warning signals. Pupation can take as little as a week in the tropics, to up to about six months in species in cold climates.

Adults visit flowers to obtain sugars to power flight. They tend to get little protein in their diet and so use proteins stored during the caterpillar stage to help make the eggs. Some moths have no mouthparts and live short lives using only stored food from the caterpillar. Pheromones are the main form of male-female attraction, with several different organs that can be everted as fluffy dispensers to waft the perfume about. The antennae of the males are extremely adept at picking up these scents, starting a search with as little as single molecules hitting the receptors. In some butterflies the wild patterns play an extra role, with males displaying their bright hues during courtship flights.

Another strategy often used is camouflage, especially by moths, which need to hide in broad daylight. Many butterfly species have leaf or other cryptic patterns on the underside of their wings, so that they disappear at rest.

And the big question... What is the difference between butterflies and moths?

Butterflies are simply several families of day-flying moths. There are some subtle differences, such as the antennae ending in a small club and not a point. Butterflies also lack a velcro-like structure on the inside wing surface, which couples the wings in flight among most moths. This is why butterfly flight tends to be slower and more erratic than the often swift and straight-flying moths. There are exceptions to all rules, however, and several families treated as moths have clubbed antennae and are day-active, although no families classified as moths have all the features of the butterfly families. Pattern and brightness are certainly no dividers – as this book will show, moths can be every bit as gaudy as butterflies. Because there are many books on butterflies, but very few on moths, this chapter will concentrate more on the moths. Quite apart from the fact that butterflies represent only about 11 per cent of this order.

Using their rough evolutionary sequence, the families will be presented from the tiny ancient caddisfly-like forms to the most advanced Noctuid moths. Butterflies will then have their own subchapter.

One of the splendid **ghost moths**, family **Hepialidae**, which are rarely glimpsed during their one day of flying life. *Aenetus scotti* is from the tropics in Australia, which is home to nearly one quarter of the world fauna of about 500 species. 5cm (2in) long.

A freshly emerged *Abantiades hydrographus*, one of the jewels among **ghost moths**, from Western Australia. Wings 12cm (5in) when spread.

The **fairy moths**, family **Adelidae**, are about 300 species of small often very ornate moths. Night-flying species are drab, but the day-flying ones have splendid metallic scales. Both belong to the genus *Nemophora*, are from Australia, and are about 0.6cm (0.2in) long.

An unidentified species of **fairy moth**, family **Adelidae**, day-active in the highland rainforests of New Guinea. The huge eyes are an adaptation of the daytime habits, which may include visual mate-finding behaviour. 1cm (0.4in) long.

The family **Psychodidae**, called **case moths** or **bag moths**, is known around the world for the elaborate homes the caterpillars create to hide and protect themselves. More than 1,000 species show great variety of case architecture. In some species the females do not develop wings after pupation and stay in the cases, where the flying males find them.

Proper **case moth** carpentry! Left: *Lepidoscia* sp. from Australia, 4.5cm (1.8in) long. Right: Log cabin from New Guinea, 1.8cm (0.7in) high.

Above: The variety of **case moth larva** case designs is endless. Left: Simply silk and dirt, in a huge species from Australia, 10cm (4in) long. Middle: One of many leaf designs, about 3.5cm (1.4in) long. Right: A snail-shell design using a seed pod and silk, from New Guinea, 1.8cm (0.7in) high. Adult case moths are shown below.

The previous page shows many designs of **case moth** larval homes. Top left is *Lepidoscia* sp., which makes the stick cases, and here a typical adult of this genus. 1.2cm (0.5in) long.

Hyalarcta nigrescens is interesting for having no scales on the wings, and being a male **case moth**, whereas the female pupates into a wingless form in the case and does not emerge. 1.2cm (0.5in) long.

The family **Tineidae** has around 3,500 species, quietly going about their business and often looking quite splendid, but the common family name '**clothes moth**' is unfortunate, as only about 10 or so species destroy our clothes (below). Left is an *Edosa* sp., 1cm (0.4in) long, and right is *Moerarchis clathrata*, from Western Australia, 1.5cm (0.6in) long.

This drab little moth and its caterpillar are the cause of much annoyance around the world. It is one of 10 or more **clothes moth** species that do attack natural fibres, especially wool. *Tineola bisselliella* is about 0.6cm (0.3in) long and does its work in dark cupboards.

The **Gracillariidae** are tiny moths with attitude. The way they stand in the wild is important for identification. Above left is a *Caloptilia* sp. from Australia, and below left is a species from New Guinea. Above right is the weirdest of them all, more like a flying fish out of water than moth. The 2,000 species are mainly **leaf miners**, their tiny caterpillars, well hidden, eating the soft tissues between leaf layers (below right).

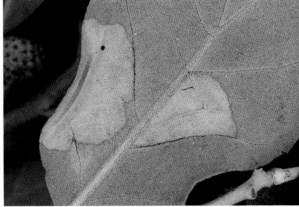

The **Yponomeutidae** is a family of about 600 species, with the habit of sitting with their wings curled into a tube. Their caterpillars often hide in silk shelters over the food leaf. Below left is *Yponomeuta* sp. and below right is *Atteva* sp., both from tropical Australia, and both about 1.2cm (0.5in) long.

The next three families are part of a major group, the **Gelechioid** families, identified by having their palps (mouthparts on either side of the proboscis), which are enlarged and curved upwards like pointed tusks, although they are covered in scales. The **Oecophoridae** are prominent in the dry habitats of Australia where many species have larvae adapted to eating dead leaves and even animal scats, including those of the koala. The **Gelechiidae** is the largest family with more than 4,500 species with diverse lifecycles, including several fruit pests. Most are on the small side, around 0.8–1cm (0.3–0.4in) long.

About 2,300 of the 3,300 species of **Oecophoridae** are found in Australia. Left: A eucalyptus feeder, *Wingia aurata*, 1.6cm (0.6in) long. Right: A new species of *Aristeis*, with unknown habits, 1cm (0.4in) long.

These **oecphorid moth** species show the variety within the family. Above left is *Pseudaegeria phlogina*. Above right is an unidentified species from islands north of Australia – note the large tusk-like palps. Below left is *Habroscopa iriodes*, and below right is another species of *Pseudaegeria*, which mimics braconid wasps. All about 1.2cm (0.5in) long.

Wingia lamertella, an **oecophorid moth** from Western Australia, another eucalypt leaf feeder. 1.5cm (0.6in) long.

The **Cosmopterygidae** is another of the **gelechioid moth** families, often using this head-down, curled-wing stance at rest. Both are in the genus *Labdia*. 1cm (0.4in) long.

The family **Gelechiidae**, the namesake of the group on this spread, is very diverse, with more than 4,500 species, and largely with unknown habits at species level. Larvae can live in fruit, seeds, between and inside leaves, or form galls. Left is *Dichomeris ochreoviridella* from Australia, and right a tiny unknown species from Indonesia.

The **wood moths**, or **goat moths**, family **Cossidae**, are about 700 species of large to very large moths. Their larvae burrow tunnels in living stems, roots and the trunks of trees. They emerge after two to three years leaving a characteristic pupal shell sticking out of the tunnel exit. Left is *Chalcidica minea* from Indonesia, 7cm (2.8in) long, and right is *Cossodes lyonetii* from Western Australia, 5.5cm (2.1in) long. Both are extreme examples in a family with largely brownish species.

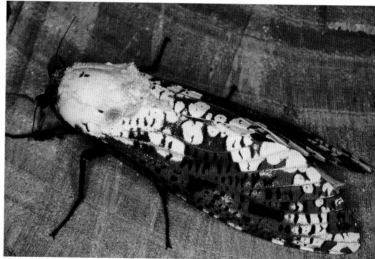

The Aboriginal people of Australia know **goat moths** very well. In the deserts it is a source of essential protein, dug out of various acacia and other tree roots. The common name **'witjuti' grub** is used nowadays for this 7.5cm (3in) long feast.

Xyleutes persona is a common rainforest **wood moth** species found from Australia to India. 6cm (2.4in) long.

Family **Tortricidae**, the **leaf-roller moths**, has more than 5,000 species. These small to medium-sized moths have larvae which usually join leaves together with silk, either flat or rolled, as they eat from inside out. Others may eat seeds, fruit or dead leaves. On the left is *Goboea opisana*, and on the right is a warning-marked *Thaumatographa* sp. from Costa Rica. Both about 1.2cm (0.5in) long.

The **Choreutidae**, known as **metalmark moths**, are a small but fascinating family of about 350 tiny species. Several genera have been shown to mimic the display behaviour of male jumping spiders when in danger of being attacked by this major predator of small moths. They open their wings in a unique way (above left) and move about in the same jerky motion as the spider, often escaping with their lives. Other species have spectacular patterns, displayed during their daytime habits, including *Choreutis periploca* (above right) and *Saptha libanota* (below left), both from Australia, and a species from Borneo (below right). All between 0.8–1cm (0.3–0.4in) long.

Several families of moths have a few species which mimic wasps for protection, but one, the **clearwing moths**, family **Sesiidae**, specialise in it. About 1,000 species often have the warning markings, imitation of a waist, and clear wings of wasps. They are day flying, but largely tropical and rarely seen as many may be at flowers in the canopy. Shown here are *Chamaesphecia* sp. (top left), *Sesia apiformis* (top right), clearwing hornet mimic *Bembecia ichneumoniformis* (below left), all from Europe, and an *Osminia* sp. (below right) from North America. All are around 2–3cm (0.8–1.2in).

The **Glyphipterigidae** is family of small day-flying pollinators, containing about 400 species. Shown here is *Glyphipterix* sp., from Australia, 1cm (0.4in) long.

The **Castniidae** is another family of day-flying moths, and have clubbed antennae like butterflies. Larvae develop on grass roots. *Synemon jcaria*, 2.5cm (1in) across.

The **Alucitidae** are known as the **many-plumed moths**, and have these remarkable wings composed of feather-like ribs. There are only about 130 rarely seen species. *Alucita* sp., 1.4cm (0.6in) across.

The **Crambidae** is a huge family of mainly night-flying moths, but all have exceptions, like this species pollinating flowers during the day in New Guinea. 2.5cm (1in) across.

The **Lacturidae** is a family of 100 species of small warning-patterned moths. *Lactura leucophthalma*, 1.2cm (0.5in) long.

The family **Lacturidae** is closely related to the burnet moths, although many species are not day flying, and most are restricted to the tropics and subtropics. *Lactura suffusa* breeds on fig trees in Australia and New Guinea. 1.2cm (0.5in) long.

Zygaena filipendulae, a **burnet moth** from central Europe. Note the variations within this and other species on these pages.

The warning patterns of **burnet moths** never cease to surprise, as they are meant to do, in order to deter predators from eating their poisonous bodies. *Zygaena laeta* from Europe, 1.4cm (0.6in) long.

Zygaena filipendulae from Europe and Asia. 1.4cm (0.6in) long.

Zygaena fausta from Europe, 1.2cm (0.5in) long.

Pollanisus cupreus from Australia, 1cm (0.4in) long.

Zygaena trifolii from Europe, 1.2cm (0.5in) long.

Extreme blue **burnet moth** from Indonesia, 1cm (0.4in) long.

Zygaena trifolii, variation, from Sweden, 1.4cm (0.5in) long.

One of the largest moth families is the **Pyralidae**, with more than 16,000 species. It has recently been tinkered with and a large number of species split into a new family the **Crambidae**, which was formerly a subfamily. Until the dust settles on these changes, the old **Pyralidae** borders will be used here. Many have a triangular shape when in sitting posture, with long thin legs. One large group, the Phycitinae roll their wings at rest. Larvae are mainly leaf associated, hiding in silk-joined shelters.

The inquisitive high stance of a typical **pyralid moth**. *Palpita pajnii* from Australia, 2.5cm (1in) wingspan.

Unidentified **pyralid moth** species from New Guinea, in a 'fighter-jet' style pose. 1.8cm (0.7in) long.

Pyralid moth from Malaysia. 3.5cm (1.4in) across.

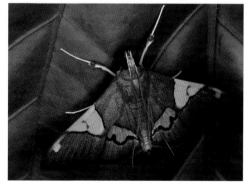

Pyralid moths in the subfamily **Nymphulinae** often have caterpillars which live underwater in silken leaf cases and shelters. Amazingly some species even walk into the water to lay their eggs. This species from New Guinea, 1cm (0.4in) long.

An unorthodox standing posture for a **pyralid moth**, with its palps forming a 'beak'. From Costa Rica, 1.8cm (0.7in) long.

The subfamily **Phycitinae** contains hundreds of species with this rolled wing sitting stance. Most are fairly drab, but this one from New Guinea stands out. 1.5cm (0.6in) long.

Glyphodes stolalis is distributed from India all the way to Fiji. Its larvae feed on fig trees. 3.5cm (1.4in) across.

This striking **pyralid moth** species, *Vitessa zemire*, found from Indonesia to Australia, is advertising warning markings to predators. 3cm (1.2in) long.

Currently the largest moth family is the **Geometridae**, with more than 21,000 species. The majority of species have a recognisable shape, spread out flat, with wings outstretched sideways, upper pair longer than bottom pair. There are of course exceptions. The caterpillars tend to be the classic 'inch-worms' or 'loopers' with their legs and pro-legs spread out to the ends of the body, causing them to bunch up in the middle as they move. They are leaf feeders in the open, however many have excellent camouflage as sticks, stems or leaf edges.

Simena sp., from Costa Rica. This blue with white stripe pattern is mimicked by many moths across several families, some of which are therefore poisonous. 3cm (1.2in) across.

Pseudopanthera macularia from Europe has day-flying habits, pollinating flowers alongside butterflies. 2.5cm (1in) across.

Comostola cedilla from Australian and New Guinea rainforests. 2cm (0.8in) across.

Geometrid moth from the highlands of New Guinea and Indonesia. 4cm (1.6in) across.

Problepsis apollinaria from New Guinea and Australia. White is very rare among insects, and silver more so. 2.5cm (1in) across.

The typical **geometrid looper** pose. These caterpillars move by stretching and bunching the body.

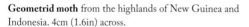

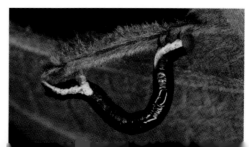

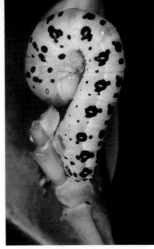

Dysphania numana is known as the **four o'clock moth**, as the adults tend to be active from late afternoon. They are found from Indonesia to Australia, and the very showy 'question mark' caterpillars feed on rainforest plants. Moth up to 8cm (3.2in) across.

Above left shows a typical **geometrid moth**, *Anisozyga fascinans*, with its fluffy pheromone release organs out, letting the scent waft on the wind. Above right is an oddly folded species from Western Australia, *Arcina fulgorigera*, 1.8cm (0.7in) long. Below left is a geometrid from Malaysia, *Plutodes malaysiana*, with cryptic patterns matching the disease circles of the leaves it sits on. Below right is *Corymica pryeri* from New Guinea.

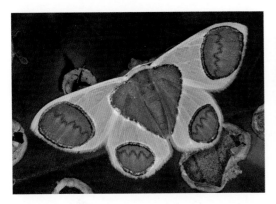

The 1,000 species of the **cup moth** family, **Limachodidae**, are named after the white silken 'cups' which house the pupae. They are better known for their remarkable wild shaped and patterned caterpillars, with names like 'chinese junk' and 'slug caterpillar'. These have many barbs and even anemone-like stingers, all of which pack a painful punch if touched.

It is the **cup moth caterpillars** which draw the most attention. All are somewhere between green and wildly warning patterned, telling all comers that the barbs, hairs and spines actually are poison tipped. They contain histamines that cause allergic reactions, and chemicals which cause stinging pain. Left from Ecuador, and right from Australia, both about 2.5cm (1in) long.

Extra barbed species from New Guinea, 3cm (1.2in) long.

The white circles of barbs are everted from hiding when the caterpillar is threatened. Just like a sea anemone, and with similar poison. *Doratifera* sp. from Australia, 2cm (0.8in) long.

A fairly typical adult **cup moth**, rather stout and densely hairy. 2.4cm (1in) across.

Anaxidia sp., from Australian rainforest, 2.4cm (1in) long.

The **Uraniidae** family has 700 species, divided into large, showy, mainly day-flying species, and small oddly shaped night species. The Uraniinae subfamily has many white, and some very big and showy butterfly-like species, most restricted to the tropics. Caterpillars live exposed on their food plant leaves.

Few butterflies compare with the beauty of some **uraniid moths**. South America is home to *Uramia* species. Here males congregate to drink needed trace elements from the river clay. 10cm (4in) across.

Urapteroides astheniata is part of a large group of species with a white, butterfly-like body, some even with little tails on the hind-wings. From New Guinea, 4.5cm (1.8in) across.

This odd way of furling the wings at rest is typical of about a third of **uraniid moths**. This is *Phazaca interrupta* from inland Australia. 2cm (0.8in) across.

Moth or butterfly? This large **uraniid moth**, *Alcides metaurus*, is from tropical eastern Australia. At times it appears in large numbers in the rainforests, where it breeds on various vines. 7cm (2.8in) across.

This very large **uraniid moth** is a species of *Lyssa*, a genus found from Australia to Indonesia. They are active at night and often come to lights. 11cm (4.4in) across.

Any insect in extreme close up takes on very new attributes. More like a yak or highland cattle than a moth, this species of **Lasiocampidae** demonstrates the extreme hairiness of this family. The light structure in the background is the wings. 4.5cm (1.8in) across.

Trabala ganesha is in the family **Lasiocampidae**. From Borneo, 4cm (1.6in) across. There are 1,500 species in this family. They have a variety of sitting postures, and this odd spread is typical.

Here a different sitting pose, typical of some **Lasiocampidae** moths, shows where the common name of **snout moths** originates. *Pararguda* sp., 2cm (0.8in) long.

The **Bombycidae** is a small family of only about 350 species of robust moths, famous for containing the **silk moth**, domesticated for the excellent silk it uses to spin its cocoons. There are many other moths, especially in the **emperor moth** family **Saturniidae**, which produce very good silk.

The Chinese **silk moth**, *Bombyx mori*, has been domesticated for so long, about 5,000 years, that it is not found in the wild at all. The fluffy adult has emerged from the rearing chamber, which below left, shows the pupa. It is the silk wound around the pupa case which is unwound to produce commercial silk, an industry known as sericulture.

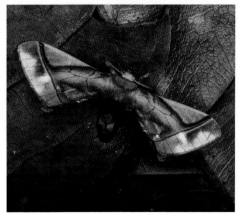

Epia sp., from Costa Rica represents the other side of the **Bombycidae**. Many species have this curled wing posture, deep rich patterns, and live in tropical forests. 3.5cm (1.4in) across.

The family **Drepanidae**, with about 650 species, has this 'hook-wing' posture among most of its species. Top left is *Astatochroa fuscimargo* from Australia. Above is a Drepanid outside the norm, a one-off wonder, *Hypsidia erythropsalis* from rainforests in eastern Australia, 3.5cm (1.4in) across.

As with the cup moths, the **Anthelidae** family is better known for its caterpillars than its handsome moths. Most are exceptionally hairy and deliver a strong histamine dose to the touch, causing allergic reactions, rashes, and serious trouble if rubbed into an eye. The 90 species are only in Australia and New Guinea. Both above are *Anthela* sp., moth 4.5cm (1.6in) across, and caterpillar 6.5cm (2.4in) long.

The **Megalopygidae** family is restricted to the Americas, where its adults are fluffy and the larvae are called puss caterpillars – an odd name as they are surely better known for having the weirdest hairdos of any moths. *Trosia* sp., adult 2.5cm (1in) long and caterpillar 4cm (1.6in) long, from Costa Rica.

The **emperor moths**, **atlas moths**, and **moon moths**, family **Saturniidae**, are large to huge showy moths with about 1,500 species. The largest moths in the world belong here, as do the spectacular moon moths. Most species have combinations of browns, yellows and golds, but fantastic markings, and large showy eye-spots abound too. The caterpillars are large, fat and bright, often with raised hair-tufted turrets. Adults do not feed and live a very short life.

Above: the caterpillar of the Australasian **Hercules Moth**, *Coscionocera hercules*, is close enough to life size.

Previous page, above: the female Australasian Hercules Moth is a contender for the title of largest moth in the world, when measured by surface area of wings – one measured 360 sq cm (56 sq in)! The wingspan is 25cm (10in). Previous page, below: caterpillar of a moth from the genus Automeris from Ecuador, 11cm (4.4in) long.

One of the jewels of the European insects, the **Spanish Moon Moth**, *Graellsia isabellae*, found in pine forests from Spain to Switzerland. 8cm (2.8in) across.

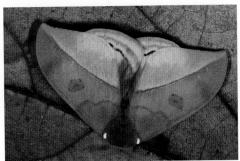

Moon moths are charaterised by long 'tails' of male lower wings. *Eudaemonia argus* from Ghana, 7cm (2.8in) across.

The **emperor moth**, *Automeris zozine*, from Ecuador, demonstrates how the eye-spots are used to startle potential predators. 7cm (2.6in) across.

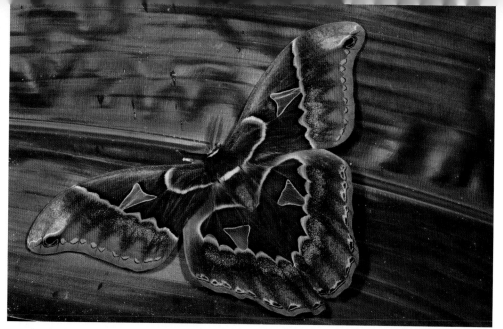

Emperor moth, *Rothschildia orizaba*, from Costa Rica, 22cm (8.5in) across.

Above: White is very rare among insects, making this exotic **emperor moth** caterpillar extra special. From Ecuador, 8cm (3.2in) long.

Opposite: One of the more dramatic **moon moth** males, *Argema mittrei* from the rainforests of eastern Madagascar, where it is called the **comet moth**. It is slow-flying, with the tails hanging down. It is 16cm (6.4in) across and long.

Actias selene, the **Indian Moon Moth**, is found from India to Indonesia. This image was taken in northern Thailand. 14cm (5.6in) across.

The **Luna Moth**, *Actias luna*, is one of the largest and most attractive moths in North America. Its fat caterpillars use a wide range of food plants, including alder, birch, elm and willow. 11cm (4.4in) wingspan.

Hawk moths, family **Sphingidae**, are very distinctive, with bodies that often look very muscular and streamlined, and indeed they are very fast fliers. There are more than 1,200 species in the world, mainly in the tropics and subtropics. Their caterpillars can be cryptic or warning patterned, sometimes with surprise eye-spots and a rearing display, but always easily identifiable by a 'horn' standing erect at the rear end.

Real hummingbirds are found in the Americas, but the **hummingbird hawk moths**, genus *Macroglossum*, live in Eurasia from Ireland to Japan. Their super-fast wing-beats while hovering make a loud hum. A long proboscis can probe for nectar in very deep flowers. Many species of this genus are day-flying. 2.5cm (1in) long.

This short-winged, fast-flying **hawk moth** species, *Angonyx papuana* from New Guinea, is 4cm (1.6in) long.

Many **hawk moth caterpillars** have bright warning markings, and most have a prominent 'horn' pointing up at the rear. From New Guinea, 8cm (3.2in) long.

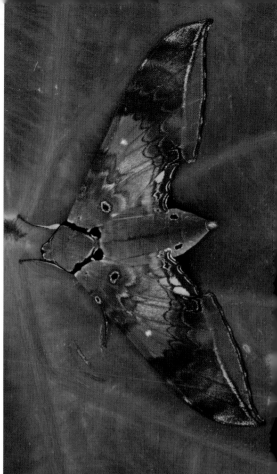

The two main sitting postures among **hawk moths** are the long-winged horizontal pose shown here and the tight triangular pose on the previous page. The two species here are large. Left is *Adhemarius dariensis* from Costa Rica, 12cm (4.8in) across. Right is *Ambulyx phalaris* from New Guinea, 14cm (5.6in) across.

Left: A large dead-leaf mimic **hawk moth** from Costa Rica, *Protambulyx strigilis*, 13cm (5.2in) across. Right: One of Europe's most handsome moths, the rare **Willowherb Hawk Moth**, *Proserpinus proserpina*, whose caterpillars feed on primrose and willowherb. 5cm (2in) across.

The **Death's Head Hawk Moth**, *Acherontia lachesis*, provided much dread for humans in ancient Europe. It is also unique in its habit of raiding bee nests to steal honey. 5cm (2in) long.

Caterpillars of the **Spurge Hawk Moth**, *Hyles euphorbiae*, are biological control heroes for eating the weed *Euphorbia esula* in Europe.

A perfect resting spot chosen by this **hawk moth** in a rain-drenched forest in Madagascar disguises it among the wet forest floor leaf litter. 3.5cm (1.4in) long.

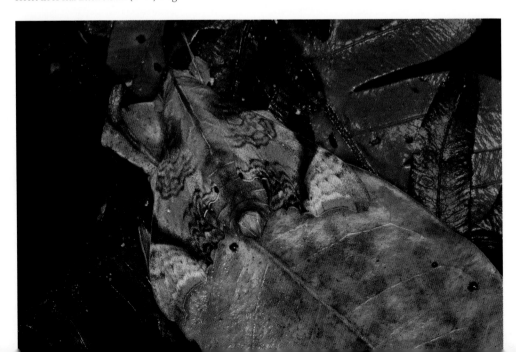

The **Notodontidae** family are medium to large moths, usually very 'hairy' and robust. Their caterpillars are interesting for having many shapes, from normal, very hairy ones, to fully bizarre shapes like the lobster caterpillar. There are around 2,800 species and the diversity is greater in warmer climates.

A typical **notodont moth** has a furry hairdo at the front and wings covered in large rough scales. This is *Cynosarga ornata* from the Australian tropics, 2cm (0.8in) long.

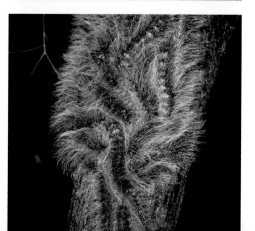

There is a story here. The **processionary moth caterpillars** are legendary in Australia. They are the communal larvae of *Ochrogaster lunifer*, also known as the **bag-shelter moths**. They live in large numbers mainly on wattles, hiding at night in huge silk bags. As they defoliate the tree, they go walk-about in long lines with heads touching bums to maintain the 'procession'. However if the leader is somehow turned, so that it winds up catching up with the rear, it can result in a circle which will move with such strong instinct only, that they eventually die in the process. Their other claim to fame are the severely irritating hairs which cause serious allergic itching and rashes. Even old hairs in the bags or from cast skins are dangerous. The exceedingly hairy adult is on the next page.

Opposite: A close-up portrait of the **bag-shelter** moth *Ochrogaster lunifer*, 3cm (1.2in) long.

There are exceptions to every rule. The rough scaled hairy look of most **notodont moths** is absent in this amazing mimic of a dead leaf. From Belize, 2cm (0.8in) long.

The **puss moth**, *Cerura multipunctata*, has a well defended caterpillar. The rear filaments are whipped about when disturbed, and formic acid can be sprayed from the front end. From Australia, 3cm (1.2in) long.

The **lobster moth caterpillar** is famous for its weirdness. *Stauropus fagi* is found across the northern hemisphere – this image is from Germany. It can arch and wriggle in a very threatening but harmless display if attacked. 6.5cm (2.6in) long.

White is rare among insects, so mostly white species are noteworthy. Left is *Epicoma zelotes*, found in Western Australia breeding on eucalyptus, 2.2cm (0.9in) long. Right is a **notodont moth** from South Africa, 4.5cm (2.8in) long.

The **banksia moth caterpillar** is not quite as wild as the lobster moth caterpillar (previous page), but it does rear its head backwards in a similar fashion. Here the head is the beige blob pointing at the stars. *Psalidostetha banksiae* from Western Australia, 7cm (2.9in) long.

Until recently the largest family of moths was the **Noctuidae**. It had tens of subfamilies and a relationship with a few other quite distinct families. Modern molecular scientists have been tinkering with these groupings, and the result is a state of flux waiting for the dust to settle on many new families, especially the **Erebidae**, which is made up of bits of older, morphologically distinct groups. The remainder of this moth chapter covers basically what used to be the **Noctuidae**, and the separate families of **Lymantriidae**, the **tussock moths**, and **Arctiidae**, the **tiger moths**. If the reader wants to pursue further information, chances are that most of it will be arranged under the old time-tested groups, and so both names will be used.

Lymantriidae, the **tussock moths**, are now a subfamily **Lymantriinae** in the new family **Erebidae**.

Even more than the notodont moths, **lymantriid moths** are very fluffy all over. Shown here is a species from New Guinea, 3cm (1.2in) long.

A **lymantriid moth** from Malaysia is unique by having 'windows' in its wings. 2.5cm (1in) across.

A **tussock moth** from Australia in close-up shows the deep layers of hair-like scales covering every part of its body. Rather like an English sheep dog, it is hard to see how it sees. *Iropoca* sp., 3cm (1.2in) long.

The name **tussock moth** actually refers to the caterpillars of the **Lymantriidae**. Their body is covered in tufts, or 'tussocks', of erect hairs. Above is *Teia athlophora* from Western Australia, and below a species from Borneo. The blue spheres are not its eyes but sensory hair bundles. Both about 4cm (1.6in) long.

The **tiger moths**, formerly family **Arctiidae**, now subfamily **Arctiinae** in the new family **Erebidae**, are usually deeply warning patterned, and some day-flying species mimic wasps. Their caterpillars tend to eat poisonous plants and store the poisons in the adult body. There are about 6,000 species, plus another 100 in the closely related former family **Aganaidae**, now the subfamily **Aganainae**.

The genus *Euchromia* contains many warning-patterned species from South-East Asia to Australia. This one from New Guinea is 4cm (1.6in) across.

Cyana malayensis is part of a group of **tiger moth** species which are imitated by other moths. From Malaysia, 2.8cm (1.1in) long.

Costa Rica is home to many species of wasp-imitating **tiger moths**. With this layer of protection, on top of warning markings, not all of these mimics need to actually be poisonous, as is common among tiger moths. *Halysidota tassellaris*, 2.5cm (1in) across.

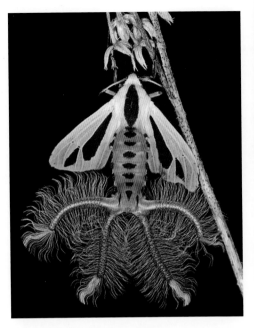

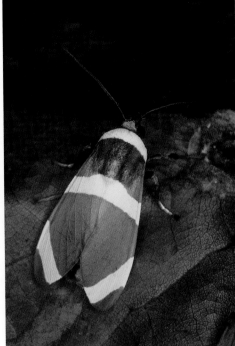

The fantastic hairy 'tails' of this moth are called coremata – structures extended by blood pressure from the last segment of male **tiger moths**. The hairs are releasing pheromones to attract females to mate – a very rare sight. *Creatonotos gangis* from Australia. Moth alone 2.5cm (1in) long.

A small striking **tiger moth** from Ecuador. *Amaxia* sp., 1.5cm (0.6in) long.

A simple but striking warning pattern on this **tiger moth** from Costa Rica. *Viviennea* sp., 3cm (1.2in) long.

The tropics do not have the monopoly on boldly patterned moths, as can be seen with this mating pair of **Wood Tigers**, *Parasemia plantaginis*, from Europe. 2cm (0.8in) long each.

Most patterns on moth wings can be related to other species in an evolutionary sequence. Sometimes though, truly original wing 'art' surprises the seasoned observer, as with this **tiger moth**, *Halysidota* sp., from Costa Rica, 3.5cm (1.4in) long.

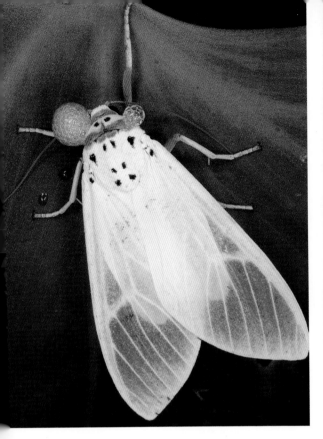

Most species in the **Aganainae** subfamily of **tiger moths** are poisonous to predators, but some also exude these spheres of acrid alkaloids just to be sure. *Amerila rubripes* from Australia and New Guinea, 3.5cm (1.4in) long.

A part of the **tiger moth** subfamily have this squat, small shape, but are still adorned with bold warning markings. From New Guinea, 1.5cm (0.6in) long.

Oeonistis delia is found from Indonesia to New Caledonia, and apparently breeds on lichen. 3cm (1.2in) long.

Ormetica sp. from Costa Rica, 3.5cm (1.6in) across.

Although not as specialised as the Sesiidae, **tiger moths** have hundreds of wasp-mimicking species, especially in South America and South-East Asia. Above is a *Cosmosoma* sp. from Ecuador, 2.5cm (1in) across. Below left and below right are species from Thailand, with the mating pair especially good as hornet mimics.

The woolly bear is a **tiger moth caterpillar** with stark patterns warning predators of poisons stored from food plants, and irritating hairs. *Pyrrharctia isabella* from North America, 4cm (1.6in).

The genus *Idalus* has many **tiger moth** species in Central and South America. This one from the Costa Rican highland forests. 2.5cm (1in) long.

Another wasp-mimicking clear-wing **tiger moth** in the genus *Cosmosoma* (see previous page). From Costa Rica, 1.8cm (0.7in).

Spotted **tiger moth** from Belize. *Hypercompe* sp., 3cm (1.2in) long.

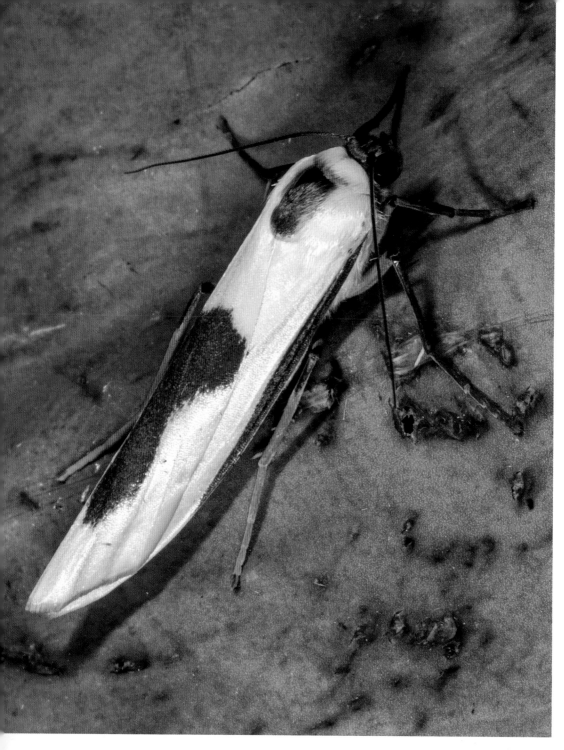

An unidentified **tiger moth** from the cloud forests of the New Guinea highlands. 2cm (0.8in) long.

The family **Nolidae** was once part of the giant **Noctuidae** family.
Around 1,400 species worldwide.

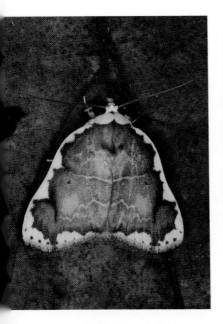
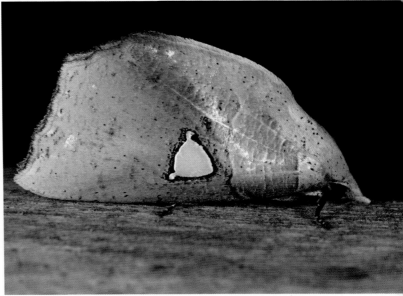

A triangular sitting posture is normal for the **Nolidae**, but there are always weird exceptions. Left is the 'typical' *Chandica quadripennis* from Malaysia, 1.5cm (0.6in) across. Right is a dead-leaf mimic from Australia, 2cm (0.8in) long.

Ariola coelisigna is another typical **nolid moth** from New Guinea and Australia, 1.5cm (0.6in) long.

A cryptic **nolid moth**, with a pattern like paint that has run in the rain, hiding in the wet leaf litter of a New Guinea highlands forest. 1.6cm (0.7in).

The previously largest moth family, the Noctuidae, has been tinkered with recently, and now many of its subfamilies are in a new giant family called **Erebidae**. In this section the old family borders are used, with acknowledgement of the new. Details of both are required in order to pursue further knowledge about these groups. What is left of the **Noctuidae** is still famous for having some of the world's worst crop pests, the armyworm, earworm and cutworm group of species. However the vast majority are harmless, and often quite stunning. Actual numbers of noctuid species are currently unclear until the dust settles on the reclassification.

One part of the **Noctuidae** which has been preserved is the subfamily **Agarisitinae**. Many species are brightly patterned, some are day-flying, and a few are pests of stone fruits and grapes. Left is the African peach moth, *Egybolis vaillantiana*, from Kenya, 3.5cm (1.4in) across. Right is an unidentified species from New Guinea, 3.5cm (1.4in) across.

These moths are not much to look at, but represent the group most feared by farmers, and indeed they are responsible for immense crop damage around the world. Top left is *Agrotis munda*, the **Brown Cutworm**. Top right is *Helicoverpa armigera*, the **Corn Earworm**. Below left is *Spodoptera mauritia*, the **Lawn Armyworm**. All are around 2cm (0.8in) long. Belwo right is the caterpillar – the damage-causing part of the lifecycle – of *Heliothis punctigera*, one of the **budworm** species. Many similar species are found around the world.

Alypophanes iridocosma from New Guinea and Australia, 2cm (0.8in) across.

This unidentified **owlet moth**, family **Noctuidae**, from Costa Rica, sits flat on top of leaves, somewhat imitating bird droppings. 2cm (0.8in) across.

The largest noctuid subfamily, the **Catocalinae**, has been moved to the new **Erebidae**. This group has medium to large species, with two very characteristic sitting postures. The sharp triangle-shape above, and below the spread showing much of the bottom wings. Top left is from New Guinea, and top right is *Avatha bubo* from Malaysia, both 3cm (1.4in) long. Above left is *Speiredonia mutabilis*, and right is *Donuca orbigera*, both from Australia and around 5cm (2in) across.

Many of the noctuid moths moved to the family **Erebidae** are camouflage experts. Above is a species from Costa Rica hiding among rich deposits of rainforest leaf litter, and below is *Epicyrtica metallica* from Australia, spending the daytime among lichen.

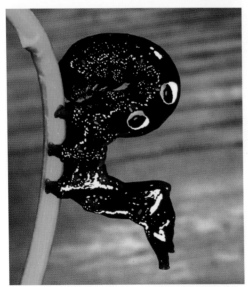

One group of moths moved from the Noctuidae to the **Erebidae** are the **fruit-piercing moths**. They reverse the moth norm, and have harmless beautiful caterpillars (below left), while the adults are adapted to pierce and damage fruit with a toughened proboscis. Top left shows *Rusicada revocans* stabbing an orange, and top right is the pretty but pesky *Eudocima aurantia*.

The final moth of this chapter is a true oddity. Noctuids have 'ears' and can pick up the ultrasound of bats and take evasive action. The male **Whistling Moth**, *Hecatesia exultans*, produces its own ultrasound calls to attract females, which can hear the calls from up to 25m (80ft) away. Australia, 2cm (0.8in) long.

Butterflies

As discussed in the moths introduction, the butterflies are not different from moths in any more serious way than different moth families are from each other. They represent seven families of the same order, Lepidoptera, and generally differ from other moth families by having a club on the end of the antennae, by not having their upper and lower wings joined during flight, and by their day-flying habits. Some of these features can be found among moths, but almost never all three together. Butterflies fit into two super family groups – the **Hesperoidea** has only the skipper family **Hesperiidae**, and **Papilionoidea** has the rest.

Hesperiidae – the skippers, with around 3,500 species.

The **Dot-collared Jemadia**, drinking at the edge of a jungle pool in Ecuador. *Jemadia pseudognetes*, 4cm (1.6in) across.

One of the jewels among **skippers**, a species of *Pirdana* from Borneo. 2cm (0.8in) long.

The majority of **skippers** are shades of brown and small, around 1.5cm (0.6in), but some of the tropical species are much larger and more showy. This is *Euschemon rafflesia* from Australia, 2.5cm (1in) long.

Large **skippers** in the genus *Mysoria*, lapping trace minerals on a mud bank in Ecuador. 2.5cm (1in) long.

The **skippers** are basically 'short and fat', and larger eyed, compared to other butterflies. Their short wings beat faster, and their flight is faster and less erratic. *Anisynta sphenosema* from Western Australia, 1.5cm (0.6in) long.

A very boldly marked *Cabirus* sp. **skipper** from the Amazon. 2.5cm (1in) across.

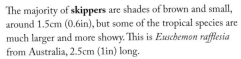

Nymphalidae – the **nymphs**, **fritillaries**, **browns**, **admirals**, **triangles** and **monarchs**, with more than 6,000 species

Admirals are a group of small to medium species, with many in the Northern Hemisphere. Above is the **Red Admiral**, *Vanessa atalanta*, from Poland, and below is *Precis orythia*, from Malaysia, in a related group called **commodores**. Both around 2.5cm (1in) across.

Above left: **Marbled White**, *Melanargia galathea*, from Europe. Above right: Red Cracker, *Hamadryas amphinome*, from Mexico – it is very blue above but red on the underside. Below left: *Bebearia senegalensis*, one of the **forester** butterflies from Kenya. Below right: One of the **clipper** butterflies, *Parthenos* sp., from New Guinea.

The **Garden Acraea**, *Acraea horta*, from South Africa. Note that the top wings are see-through. 4.5cm (1.8in) across.

The **Malachite**, *Siproeta stelenes*, from Ecuador, 4cm (1.6in) high.

Diaethria clymena, known as the **Common Eighty-eight** for its number-like markings. Central and South America, 2cm (0.8in) long.

Red-spotted Purple, *Limenitis arthemis*, from North America, 3.5cm (1.4in) high.

There are more than 300 species of **glasswings** in the Americas. They are beautiful and elusive and their wing-beats make them appear and disappear in succession. This is a species from the genus *Greta* from Mexico. 3cm (1.2in) wing length.

Another **glasswing** species, in the genus *Hypoleria*, in Costa Rica. 3cm (1.2in) wing length.

A related group of butterflies are known as the **clearwings**. Shown here is a species from the genus *Cithaerias* feeding on rotting fruit in Costa Rica. 3.5cm (1.4in) long.

The **Heliconiinae** is a subfamily of the **Nymphalidae**, with about 600 species mainly in the Americas. Most have an open-winged stance, and elongate to very elongate forewings. Above is *Heliconius eleuchia* from Colombia, and below is *Heliconius erato*. This group plays extreme mimicry games, with *H.erato* having 28 subspecies, matched by 28 similar mimics within the *H.melpomene* species. Both around 4cm (1.6in) across.

Perhaps the most famous species in the family **Nymphalidae** is the **Monarch** or **Wanderer**, *Danaus plexippus*. Found in North America and parts of the Pacific, it undertakes huge autumn migrations, for example from southern Canada to Mexico. The caterpillar (right) eats only milkweed, and stores poisons from it for protection.

One of the more striking European butterflies is the **Peacock**, *Inachis io*. At rest the wings are held upright hiding the bright eyes, but if threatened it suddenly opens the wings to startle the intruder, while making a hissing sound.

Among the largest nymphalid butterflies are the **tree nymphs**, genus *Idea*. With wingspans up to 16cm (6.5in) they fly slowly, often gliding above or through the forests of tropical Asia.

The **Red Lacewing**, *Cethosia cydippe*, is part of a group of 20 species from Australia to India. 6cm (2.4in) across.

Polygonia interrogationis is the **Question Mark** butterfly of North America, named for a tiny silver question mark near the middle of the hindwing.

The **morphos** are considered the most beautiful butterflies in the world. To see one flying along a sunlit rainforest stream is magic. When they land they disappear by hiding the electric blue inside brown wings (left), *Morpho* sp., 10cm (4in) across. Several species across the American tropics are almost never seen perched with wings open, hence the museum specimen (right) of *Morpho menelaus*, to show the amazing difference.

The **Nymphalidae** are known as 'the **browns**' and many species look similar to this general pattern. *Magneuptychia* sp., from Costa Rica.

One of the **'earl' butterflies**, *Tanaecia* sp., from Borneo, which are often seen feeding on fallen fruit on the forest floor.

Pieridae – the **whites**, **yellows**, **jezebels**, **orange-tips**, around 1,100 species.

The **Orange-tip**, *Anthocaris cardamines*, found from Europe to Japan.

The **Common Jezebel**, *Delias nigrina*, from Australia.

Apart from white, the **pierids** are known for their deep yellows, typified by the 70 species of the genus *Eurema*. This one is from Malaysia.

Small White, *Pieris rapae* – one of the two 'cabbage white' species which damage leafy vegetable crops around the world.

The **Purple-tip**, *Colotis ione*, from Kenya – a beautiful odd one out in the orange-tip group.

One of the whitest whites, the **Wood White**, *Leptidea sinapis*, from Europe even has white eyes.

The **Great Orange-tip**, *Hebomoia glaucippe*, occurs from India to Japan. 4.5cm (1.4in) wingspan.

Many butterflies, but especially male **pierids**, congregate at wet spots where they can find salts and other nutrients. This is called 'mud-puddling'. Shown here is *Catopsilia pomona* in Indonesia.

Lycaenidae – the **blues**, including the **coppers**, **hairstreaks**, **sapphires**, **oakblues** and **harvesters**, with about 6,000 species.

Some of the blues are known as **coppers** and really do have a magnificent copper sheen, such as this *Aloeides* sp. from Ghana.

The European **Common Blue**, *Polyommatus icarus*, is deep blue on the upperwing, but like many **lycaenids** it has more cryptic patterns on the underside so that it can hide when perched.

This white **hairstreak**, *Hypokopelates* sp., is one of the lycaenids with males sporting beautiful but very delicate tails. In all but just-hatched individuals, these tails are lost in flight within their forest habitats. From Ghana, 3.3cm (1.4in) long with tails.

The blues of **Lycaenidae** butterflies are usually on the upperside of the wings. The underside is either dull cryptic or boldly patterned, as in this **Purple Sapphire**, *Heliophorus epicles*, from Malaysia.

This **blue** in the genus *Zizina* perfectly illustrates the curled proboscis of butterflies. A moth proboscis is the same structure, but often partly covered in scales.

The **Brown Hairstreak**, *Thecla betulae*, of Europe and western Asia feeds exclusively on *Prunus*.

Many male butterflies seek salts and trace elements in wet soils, fruit and animal droppings. Here the **Glistening Cerulean**, *Jamides elpis*, laps up fresh bird poo in Borneo.

One of the brighter dwellers in the huge shady dipterocarp rainforests of Borneo, this member of the genus *Jacoona* is 2.5cm (1in) long with tails.

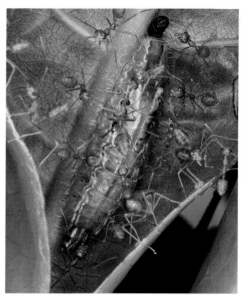

Callophrys rubi, the **Green Hairstreak**, is one of very few green species in the blues family. It is found across Europe and temperate Asia. 2cm (0.8in) across.

Common Oakblue, *Arhopala micale*, in Australia. This is one of many **lycaenid** species with caterpillars attended by ants. They produce scents and sugars which the ants love, and in return protect the caterpillars, even taking them into their nest for the night.

Like the 'push-me-pull-you' llama in Dr Doolittle stories, this butterfly, the **Long-banded Silverline**, *Spindasis lohita*, appears to have two heads. It can survive having the rear part of the wings ripped by a predator, so the larger bolder head and antennae false signal is there.

The common name of *Lycaena vigraureae* is **Scarce Copper**. This rare, boldest of the 'blues' in Europe, takes the copper brightness to extremes. 2.2cm (0.9in) across.

A **line-blue**, *Nacaduba* sp., from Western Australia.

Sometimes the most subtle shades are the most beautiful. This is the **Dusky Blue**, *Candalides acasta*, from Western Australia.

The genus *Lycaena* was named by the first modern taxonomist, Linnaeus, in 1761. The first species to be named in the genus was *Lycaena helle*, the **Violet Copper**, from Europe.

Papilionidae – the **swallowtails**, **apollos** and **birdwings**, with only around 600 species.

Probably the most familiar **papilionid** is *Papilio machaon*, the **Common Swallowtail**, found across the northern hemisphere. Its pretty caterpillar has a very wide choice of food plants. Adult 5.5cm (2.2in) across and caterpillar about 5cm (2in) long.

The genus *Papilio* has more than 100 large species, many iconic for their region, such as *Papilio memnon*, the **Great Mormon**, which is found from India to Indonesia. 12cm (4.8in) across.

Left are male **Rothschild's swordtails**, *Protesilaus earis*, drinking in salts and minerals from the muddy edge of an Amazonian creek. Right is one of the 'triangle' group, the **Green Triangle** or **Tailed Jay**, *Graphium agamemnon*, found from India to Australia.

The birdwing butterflies are huge, with very showy males. Among them is the largest butterfly in the world, and it and some others in this group of 34 species, are threatened with extinction in the wild. Above is the **Green Birdwing**, *Ornithoptera priamus*, from New Guinea and Australia, 12.5cm (5in) across. Below is a male **Rajah Brooke's Birdwing**, *Trogonoptera brookiana*, which is in decline due to deforestation in Malaysia and Indonesia, 13cm (5.2in) across.

The **Five-bar Swordtail**, *Graphium antiphates*, found from India to Indonesia, is another in the 'triangle' genus *Graphium*, which has more than 100 species.

Orchard Butterflies, *Papilio aegeus*, caught mid-flight in a nuptial dance. The male is above. Australia.

Not the prettiest swallowtail, but one with a great name – *Cressida cressida* from Australia is known as the **Big Greasy** owing to its see-through oily appearance.

South America has the morpho butterflies and Australia has the **Ulysses Swallowtail**, *Papilio ulysses*, which is dazzling blue as it flies and with cryptic dark patterns on the underside at rest. 10cm (4in) across.

This final family of butterflies is one of the most interesting. The **Riodinidae** are known as **metalmarks**. They are very hard to describe as they have species which look like classic members of all the other families. The closest relatives are the blues, **Lycaenidae**, but males have only half-length, partly atrophied front legs. About 90 per cent of the 1,500 species are restricted to South America.

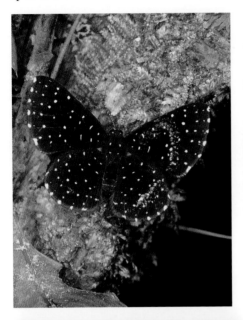

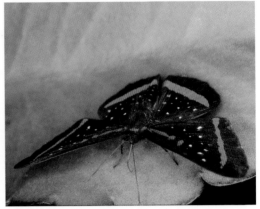

Apparently the name **metalmark** comes from the species which have many silvery, metallic spots. Left is the **Starry-night Metalmark**, *Echydna punctata*, 1.5cm (0.6in) across. Above is *Amarynthis meneria*, 2.5cm (1in) across. Both from Ecuador.

Mesosemia messeis is a member of the **riodidnid** family from Costa Rica, 2.3cm (1in) across.

A magnificent tailed **riodinid**, known as the 'blue doctor', *Rhetus periander*. From Brazil, where it displays in sunny patches of streams.

Laxita is one of the few genera of **metalmarks** found outside of South America. This one from Borneo is 2.5cm (1in) across.

Charis gynaea from the cloud forests in Costa Rica, 1.4cm (0.6in) across.

In keeping with this family 'borrowing' features from the other families, *Saribia* sp. metalmarks look like members of the 'browns' family (Nymphalidae), albeit very handsome versions. From Madagascar, 3cm (1.2in) long.

WASPS, SAWFLIES, ANTS and BEES

Order HYMENOPTERA

120,000 species in about 100 families

The wasp order represents an interesting window on insect evolution. It is divided into two suborders, with about 10 per cent in the ancient group, the **Symphyta**, or sawflies. These wasps lack the 'waist' which defines most other wasps, and almost all are vegetarians instead of parasites and predators. Their larvae are very caterpillar-like and live on the vegetation they eat, usually in protective groups that can regurgitate food and unpleasant chemicals if attacked. Their name derives from a saw-edged ovipositor (egg-laying tool), which is used to saw into plant tissue to lay eggs out of harm's way. Sawflies have been around for about 250 million years.

The main suborder is the **Apocrita**, with all the parasitic wasps, the ants and bees. The majority have a 'waist' formed from an extended first segment of the abdomen. Just over half of the families are parasitic, laying their eggs in other insects' larval and adult stages, or the pupa, or even the eggs. Some of the egg parasites are the smallest insects on earth, almost truly invisible at 0.17mm (0.008in)! The largest wasps are hunting species in the **Scoliidae**, at up to 6cm (2.4in) long. Because most parasitic species eventually kill the host, and never transfer to more than one host, they are not strictly true parasites, and are referred to as 'parasitoid'. Most leave the host as it is dying and pupate outside, often in silk spun cocoons.

The last 40 or so of the most recently evolved families differ by having the egg-laying ovipositor adapted (not in all species) into a defensive venom-laced stinger. This adaptation has allowed many species to form social groups which defend the nest, and in some cases allow it to become a hive that spans many generations. The ants and some of the bees and paper wasps are main examples, with the extra adaptation of separate castes. Other families, such as the hunting wasps of the **Sphecidae**, remain solitary. The paper wasps are called eusocial, with a queen and usually single-season communal nests. The hornets are a type of paper wasp too. These groups have habits that include both hunting and killing prey, and parasitoid behaviour.

Flowering plants have only been around for less than 135 million years. As they went through a species explosion in the Cretaceous period, around 100 million years ago, insects which feed on their sugar-rich products while pollinating the flowers also speciated widely. The bees are the result of this association, with more than 20,000 species in seven families, some living in social or semi-social systems, although most live solitary lives. The domesticated Honey Bee, *Apis mellifera*, carried by humans all over the world, is just one species.

Suborder **Symphyta**, the sawflies, with around 10,000 species.

Pergidae is the largest family of **sawflies**. The genus *Perga* has communal larvae called 'spitfires', here on eucalyptus in Australia, defending their group by writhing and exuding foul oil concentrates from their food. Adult 2cm (0.8in) and larvae 3cm (1.2in) long.

The **Club-horned Sawfly**, *Abia sericea*, lives in Europe and feeds on fennel plants. 1.2cm (0.5in) long.

This metallic blue **sawfly**, *Trichorachus* sp., family **Argidae**, shows its complex everted mouthparts which are used to gather nectar. From Australia, 1.5cm (0.6in) long.

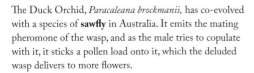

The Duck Orchid, *Paracaleana brockmanii,* has co-evolved with a species of **sawfly** in Australia. It emits the mating pheromone of the wasp, and as the male tries to copulate with it, it sticks a pollen load onto it, which the deluded wasp delivers to more flowers.

Mother **sawfly**. The eggs were hidden inside the leaf, inserted using the female's 'saw' ovipositor, and the young are protected until their own chemical defences are stronger. 1.8cm (0.7in) long.

Suborder **Apocrita**, starting with the **Parasitica** group, contains the main parasitic wasp families, with about 50,000 species. The majority of species are small to minute, but two families stand out because of their medium, noticeable size, and more than 40,000 species – the **Ichneumonidae** and the **Braconidae**. The word parasite conjures unpleasant images for humans, but all these wasps are part of the huge and essential numbers game among insects only. Parasites drive evolution and the composition of ecosystems. Without them we would be knee-deep in other insects, including our most dreaded pests.

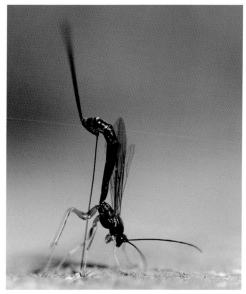

Eye-catching **braconid wasp**, *Chaoilta* sp., from South-East Asia. They parasitise jewel beetle larvae inside their wood tunnels.

Ichneumon wasps are common visitors to flowers. Most adult flying insects, no matter what their 'day-job' is, will stop at fresh flowers to replenish the sugars used to fuel flight muscles.

European **wood wasp** with ovipositor drilling into wood in search of beetle larvae to parasitise. Note the ovipositor sheath, pointing upwards. **Ichneumonidae**, 2cm (0.8in) not including sheath.

Members of a related family, the **Aulacidae**, specialise in inserting their long ovipositors into wood cracks, searching for beetle larvae to parasitise.

Members of the large **Braconidae** genus *Callibracon* are specialists of wood-boring beetle larvae hosts. 1.4cm (0.6in) long.

Tiny **Braconid wasps** emerging from silk spun pupae, attached to the body of the caterpillar they have eaten from the inside out.

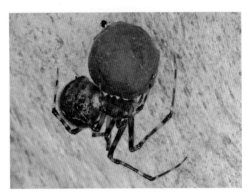

A micro **braconid wasp**, from a group which includes gall-makers and beetle-larva parasites, 0.3cm (0.12in).

A look at scale. This small window spider has parasitic **wasps** of the family **Eurytomidae** emerging from its egg sac. Wasp 1mm (0.04in).

The **Gasteruptiidae** family has wasps with the abdomen mounted and held high. More than 500 species attack bee and solitary wasp nests, where their larvae eat the eggs and larvae. Left is *Pseudofoenus* sp., 1.5cm (0.6in), and right is *Gasteruption* sp., 2cm (0.8in) long.

Most of the other **Parasitica** group of families have small to minute species, although some have very long ovipositors, adapted to reach well-hidden hosts. The following pages take a brief look at these important insect-control specialists, which are in about 35 families within the **Proctotrupoidea** and **Chalcidoidea** super families.

The **Megalyridae** family holds the record in terms of ovipositors, with some being up to eight times the length of the body! Top is *Megalyra* sp., 0.4cm (0.15in) not including ovipositor. Above is a wasp in the family **Scelionidae**, which parasitises grasshopper eggs buried in the ground. 0.3cm (0.1in) body only.

Members of the **Eucharitidae** are distinctive by having a very long and amazingly thin petiole 'waist'. They lay eggs on plants from which the hatchling larvae attach to ants and get taken to the nest, where they parasitise the ant larvae. Left 0.4cm (0.16in) and right 0.2cm (0.08in).

A parasite of ants from the family **Eucharitidae**, 0.3cm (0.12in).

Hemiptarsenus sp. from the family **Eulophidae** – a parasite of moth and fly larvae tunneling in leaves. 0.3cm (0.12in).

Members of the family **Encyrtidae** are hopper and aphid parasites, 0.3cm (0.12in).

Pycnetron sp. from the family **Pteromalidae**, a parasite of weevils. 0.3cm (0.12in).

Diaulomorpha sp., family **Eulophidae**, a parasite of leaf-mining moth caterpillars. 0.2cm (0.08in).

The 1,000 species in the family **Eupelmidae** have diverse habits, being parasites of beetles, bugs, mantids and more. 0.4cm (0.16in).

Semiotellus sp. in the family **Pteromalidae**. This genus is widely used in biological control, especially of midge fly crop pests. 0.2cm (0.08in)

Podagrion sp., family **Torymidae**, is a parasite of mantid egg cases. The bizarre rear-leg serrated expansion is a feature of this family. 0.4cm (0.16in).

Figs are an odd fruit, as the flowers are on the inside and require a very specialised pollination process. This is managed by the **fig wasps**, family **Agaonidae**. The pre-fruit has an opening for the female wasp (left) to enter and pollinate the developing flowers with pollen from her birth fig. She lays eggs and dies inside. Larvae feed on the fig and pupate inside. The bizarre devolved males (right), with only four legs and no wings, mate with new females and then dig a tunnel for the females to escape. The males die straight after. Other wasps eat the insides of figs or parasitise fig wasps, but are not part of this finely evolved mutualism. The bottom pic shows a mob of *Apocrypta* sp., family **Pteromalidae**, probing with very long ovipositors to parasitise fig wasps from outside.

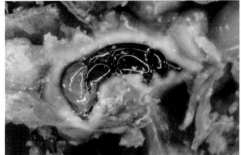

Trissolcus basalis, family **Platygasteridae**, is an egg parasite used in the bio-control of the green vegetable bug, a worldwide crop pest. Here hatching from now dead eggs. 0.15cm (0.06in).

The **Chalcidae**, characterised by enlarged back legs and black/yellow markings, are parasites of moth and butterfly larvae, usually emerging as the caterpillars start to pupate. *Brachymera* sp., 0.7cm (0.3in).

The family **Torimidae** has species which 'parasitise' plants rather insects. They inject chemicals which cause plants to form a gall, a chamber of cells, inside which the wasp larvae safely consumes the contents from inside out. From New Guinea, 1cm (0.4in) body.

The family **Chrisididae**, with 3,000 species, contains some of the most beautiful wasps in the world. Common names like **emerald wasp** and **jewel wasp** abound, but the overall name is **cuckoo wasp**, due to their habit of invading the nests of other wasps, especially the mud wasps, sphecid wasps, bees and sawflies. Their offspring eat the eggs and larvae and the provided brood food. Species which invade the nests of the more aggressive wasps can fold up into a very tough defensive ball.

A tiny **emerald wasp** on the outside of an old mud wasp nest it may have hatched from. The deep metallic tones are created in pits with many layers of differently reflecting crystal-like structures. From Madagascar, 0.5cm (0.2in).

The **Ruby-tailed Wasp**, *Chrysis ignita*, is a parasite of mason bees, solitary bees with multiple single-larva nests. They are found from across Europe and Asia, from Britain through Russia to Japan. 0.8cm (0.3in).

Stilbum cyanurum is one the largest **cuckoo wasps**, at 2cm (0.8in) long. It parasitises large mud wasp nests.

Another tiny species of **emerald wasp** from Madagascar, 0.4cm (0.16in) long.

A tiny **emerald wasp** of the genus *Chrysis* from the Canary Islands. 0.6cm (0.3in).

A boldly marked European **emerald wasp**, *Chrysis scutellaris*, 0.8cm (0.3in) long.

An **emerald wasp** from Australia. Many **chrysidids** live in dry places, as these are also frequented by their host mud wasps. *Stilbum* sp., 1.5cm (0.6in).

This **emerald wasp**, *Parnopes grandior*, in Poland is larger than most at 1.2cm (0.5in) long. It 'cuckoos' the nests of sand wasps.

The superfamily **Vespoidea** contains most of the large, showy, well-known wasps. These include the **mud wasps**, **paper wasps**, **hornets** and **spider wasps**, in about 12 families. Most have combinations of yellow and black to brown, and many possess stings. They are hunters, parasitoids and pollinators. Ants are so closely related to these wasps, that they are often classified in this group. Without the ants, there are about 15,000 species in the Vespoidea.

The **spider wasps**, family **Pompylidae**, with about 5,000 species, parasitise spiders of all sizes.

Spider wasps are medium to large and very strong. They immobilise spiders with a venomous sting and then fly or drag them to a hole where they are buried and eaten by the wasp larva. If the spider is too big to carry, the wasp cuts off the less tasty bits such as the legs.
Above: Dragging in Tasmania, wasp 2.2cm (0.9in) long.
Below: Dismembering in Borneo, wasp 3cm (1.2in) long.

Above: The giant **spider wasp** of Australia, *Cryptocheilus* sp., 4cm (1.6in) long.

Beautiful metallic **spider wasp** from Ecuador. Note the bright orange antennae. When attacking a spider, it vibrates the antennae in a display that seems to distract the spider and prevent it from mounting a better defence.

One of the **spider wasps** that specialises in hunting tarantulas in Costa Rica, 3.5cm (1.2in) long.

The **flower wasps** are in the family **Thynnidae** (formerly **Tiphiidae**), with small wingless females and larger winged males. The large **hairy flower wasps** are in the family **Scoliidae**.

The mating ritual of **flower wasps** has the flying male pick up the tiny wingless female and fly her around to many flowers while they mate. This lets her get a good feed of sugars to help make the eggs. When alone, females are ground- and bush-dwelling as they search for the larvae of scarbaeid beetles to parasitise. 2.5cm (1in) and 0.8cm (0.3in) long.

A wingless female **flower wasp** sitting in a prominent position, waiting for a winged male to pick her up on the nuptial flight.

A large winged male **flower wasp**, *Thynnus* sp., cold with morning dew. From Australia, 3cm (1.2in) long.

The variety of sizes and patterns among the male-female pairings of flower wasps makes it hard to identify species from one gender. They vary from identical patterns (previous page) to similar (left) to completely different (middle and right). The large size difference is normal. In the pair below the male is 3cm and the female only 0.6cm (1.2 and 0.2in) long.

The odd one out. This female **flower wasp** from the genus *Hemithynnus* is not tiny, and is armed with serious mandibles. 1.6cm (0.6in) long.

Members of the related family **Scoliidae** are known as the **hairy flower wasps**, with many large, hairy and big-jawed species. *Guerinius* sp. from Australia, 3cm (1.2in) long.

Members of the **Mutilidae** are known as **velvet ants**. This is another family where the males are winged and the females wingless, ground-dwelling and heavily sclerotised. They invade the nests of mainly solitary bees and parasitise the larvae inside. There are about 3,000 species around the world, and they are also known for the very painful defensive sting of the deeply warning-patterned females.

Female **velvet ant** showing the heavily sclerotised body. It needs this protection while invading the nests of solitary bees. *Ephutomorpha formicaria* from the Australian desert, 1.4cm (0.6in).

Velvet ant female from Madagascar, 1.2cm (0.5in) long.

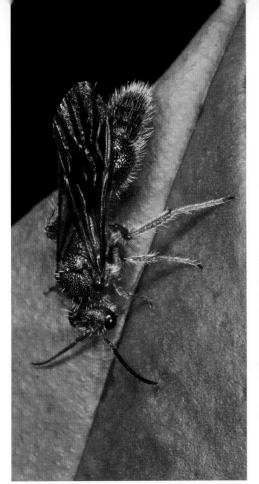

Above is a winged male **velvet ant**, 1.8cm (0.7in) long. Above right is a wingless female, 1.4cm (0.6in) long. Below is another well marked female, 1.2cm (0.5in) long. All are from the New Guinea Highlands, an area of deep patches of rainforest where amazing new insects are still being discovered.

Science fiction prop makers could learn from this spiky **velvet ant** female from an Ecuadorian highland forest. 1.5cm (0.6in) long.

The family **Vespidae** contains many of the iconic, classic wasps in five or six subfamilies. The **Polistinae** are the **paper wasps** and the **Vespinae** the **hornets**. Both these groups have fully eusocial lives, with one reproductive queen and a number of non-reproductive females living in paper nests. In Europe the nests die in the winter and a hibernating queen restarts a nest in the following season. In the tropics nests are continuous and attain huge sizes. The other main group are the **potter wasps**, or **mason wasps**, subfamily **Eumeninae** which are not eusocial and make mud nests for their young.

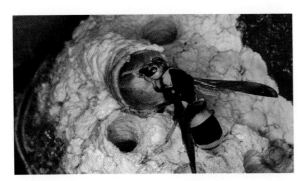

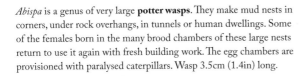

Abispa is a genus of very large **potter wasps**. They make mud nests in corners, under rock overhangs, in tunnels or human dwellings. Some of the females born in the many brood chambers of these large nests return to use it again with fresh building work. The egg chambers are provisioned with paralysed caterpillars. Wasp 3.5cm (1.4in) long.

Above right: Most **potter wasps** make either pot-shaped nests or corner-hugging mud blobs, but *Eustenogaster calyptodoma* from Malaysia is a true artist.

Delta latreillei is a **potter wasp** needing water and clay to make its mud nests. In the Australian desert, this need has them mobbing a last wet sand patch. *Delta* is a large genus found in Europe, Asia, Africa and Australia. Most species are large and have a characteristic long waist. The mud nests are provisoned with caterpillars for its larvae. 3.5cm (1.4in) long.

As the female *Abispa ephippium* **potter wasp** came to gather water for the mud mix, the male arrived and literally swept her off her feet, and into a nuptial flight.

The typical pot-shaped nests which give **potter wasps** in the family **Eumeninae** their name. Houses, especially if raised off the ground, are major habitats for potter wasps around the world. This image was taken in Malaysia.

The **paper wasps**, family **Polistinae**, make multi-layered nests from a waxy paper-like pulp. Left is one of the biggest and scariest nests, 3m (10 feet) long – it houses many thousands of the 2.5cm (1in) *Synoeca septentrionalis* wasps, called 'guitariera' for the loud buzzing noise the swarms make. Their sting is classified as the most painful of all wasps, and may require medical help. Trails in Central and South American jungles often meander strangely, to avoid their nests.

From little things... The **paper wasp** above is a queen who has just started a new colony. The brood cells contain her larvae and caterpillars for food. When they hatch she will have worker wasps to grow the nest. Wasp 1.5cm (0.6in) long.

Some **paper wasp** nests are covered (above right) but inside they are like this open nest, made of flat layers of brood cells, housing larvae. *Ropalidia kurandae* from Australia, 1cm (0.4in) each.

Polistes dominula, the common **European Paper Wasp**, standing on the water tension of a bird bath, gathering water to mix up the paper pulp for nest building – in Australia. This very adaptive species has found its way to North America, Africa and now Australia. Below left is the same species, in Europe, cleverly utilising an old Honey Bee hive box to hide the paper nest in. 2cm (0.8in) long.

The giant **paper wasp** of Australia, *Polistes schach*, 2.6cm (1in) long and packing a potent sting.

The final group of the **Vespidae** are the biggest and baddest – the **Vespinae**, the hornets and yellow-jackets, with about 80 mainly Northern Hemisphere species.

Vespula germanica, the **German Wasp**, now accidentally exported around the world. A very adaptable species eating not just insects, but also fruit, dead creatures and human cakes and garbage. In Europe its nests restart every spring, but those in Australia do not slow down, reaching 20,000 or more individuals of several castes. The paper nests are often underground, but increasingly they make use of human habitation. 1.5cm (0.6in) long.

Left: *Vespa crabro*, the **European Hornet**, the largest hornet species in Europe, successfully hunting bees, 3.5cm (1.4in) long. Above: *Vespula squamosa*, the **Southern Yellowjacket** of North America, at the entrance to the sometimes huge nests with 400,000 plus wasps, 2cm (0.8in) long.

The largest of them all is the **Asian Giant Hornet**, *Vespa mandarinia*, here drinking water for the making of paper pulp. 3cm (1.2in) long.

Large hanging paper nest of *Vespa tropica*, one of very few tropical **hornet** species, from South-East Asia. It hunts paper wasps.

The family **Sphecidae** has diverse wasps, the sand wasps, thread-waisted wasps and mud wasps. A large group called the hunting wasps, has been moved to their own family called **Crabronidae**. Both groups are parasitoid wasps stinging prey to enclose in nests and burrows, where the larvae eat these hosts.

The genus *Philanthus*, in the family **Crabronidae**, contains species known as **bee-wolves** because many of these European wasps specialise in hunting bees. Left: *P.coronatus* returning with prey to its hole in a loam bank. Right: *P.triangulum* with a Honey Bee too large to fly with, so it is dragging it back to its nest. Both about 1.2cm (0.5in) long.

The **digger wasps**, family **Sphecidae**, often have these extremely long and thin waists. Left is *Ammophila* sp., a caterpillar hunter from Australia. Right is *Prionyx kirbii*, a grasshopper hunter from Africa. Both about 2.5cm (1in) long.

The **hunting wasps**, family **Crabronidae**, are solitary hunters, catching insects to put into burrows where the larvae will grow. Left is *Cerceris* sp., hovering above its hole in Australia, and right is *Hopsiloides vespoides*, bringing a membracid hopper into its burrow in Costa Rica. They use visual spatial memory to find their minute holes after a hunting flight.

Tool-use is not restricted to crows and primates. On the left **hunting wasps** use 'boulders' to tamp down tracks around the burrow, and to block it. The wasp on the right is a member of the genus *Isodontia*, using a stick to investigate the pipe for a nest for its cricket prey.

Sand wasps. The worldwide genus *Bembix* has more than 380 species of hunting wasps with the skill to dig nests in sand that do not collapse. They provision the burrows with various insects for their larvae. Left is *B.moma* and right is *B.mareeba* showing the digging forelegs.

Ants, family **Formicidae**, are the most numerous insects on the planet, using every habitat from extreme desert, to rainforests, to our homes. There are around 13,000 named species. Their success is due to a complex, caste-based, co-operative social structure, and sophisticated methods of modifying their habitats in all manner of nests. These can be from a few ants in a glued leaf, to millions of ants in underground palaces. Combined with a sophisticated chemical 'communication', their colonies are often thought of as a single functioning 'superorganism'. All food sources, from fungus, to leaves, to seeds, to hunting, are practiced, as are slave-making of other species in order to have more workers. Defences are varied too, from brute strength of the larger soldier castes, to spraying formic acid and other chemicals, and very painful stings.

Probably the best-known ant in Europe, the **Red Wood Ant**, *Formica rufa*. Their mound nests can be several metres high, and house 400,000 ants. Foraging ants are quick to attack if disturbed, and bite and spray formic acid, which adds to the pain. 0.8cm (0.3in) long.

The genus *Myrmecia* contains about 100 species of **bull ants** or **jack-jumper ants** in Australia. It is yet another dangerous creature on that continent, as their potent sting can be fatal to people with allergies, and produces severe pain in others. The menacing jaws just hold on, while its tail-sting is brought forward. 2cm (0.8in) long.

A **bull ant**, *Myrmecia mandibularis*, from Western Australia. When their nest is disturbed they chase and sting the intruder, although they are more timid away from the nest. 2cm (0.8in) long.

The **Green-headed Ant**, *Rhytidoponera metallica*, is one of the most common ants in Australia, where it forages for everything from carrion to flowers. 0.8cm (0.3in) long.

The **snap-jaw ants** possess a ratchet-like mechanism that pulls back special elastic muscles and locks the jaws at wide open. When released, the jaws shut at 1/20,000th of a second, the fastest in the Animal Kingdom, moving at up to 230kph (145mph). The shock wave from the shutting force of 100,000 gravities can liquefy the contents of insect prey. Both species of *Odontomachus*, from New Guinea, 1.5cm (0.6in) long.

One of the largest **snap-jaw ants**, *Daceton armigerum*, from South America. See opposite for their remarkable story. 2.2cm (0.9in) long.

Green Tree Ants, *Oecophylla smaragdina*, make leaf nests glued with silk produced by the larvae, used like a glue-gun tool by the workers (above right). When nest building the ants form scaffolds that bring the leaves together under tension (above left). They are fierce nest defenders and bite in hordes, holding on even after the body is plucked from the head. Their soft green rear ends have ascorbic acid rather than the typical formic acid, and are therefore quite tasty food for humans. 0.8cm (0.3in) long.

Because the Green Tree Ants are so fierce, there are many insects and spiders that mimic them for protection and stealth. On the left is a crab spider in Australia which stalks and eats the ants, in the middle is a jumping spider in Sri Lanka, and on the right is a harmless seed bug from New Guinea.

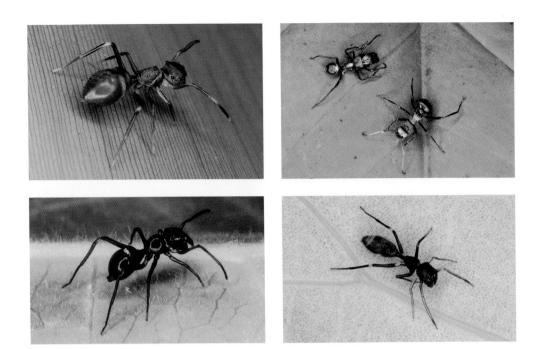

Following on from the subject of **ant mimics** (previous page), there are many hundreds of other general examples. The masters tend to be the jumping spiders, family Salticidae, as many have adapted to carrying their front pair of legs (one pair too many) in a manner just like the elbowed antennae of ants. Top left is a jumping spider from Malaysia, above left is a *Myrmarachne* sp. of jumping spider from Australia, top right is the model ant and the mimic spider meeting, and below right is a spider perfectly mimicking a local ant in Malaysia.

A splash of bright red is not common among **ant** species. Left is an unidentified species from a tropical mountain in Australia, and right is a *Crematogaster* sp. from Borneo. 0.8cm (0.3in) and 1cm (0.4in) long.

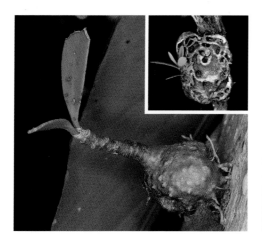

Some ants have co-evolved a mutualistic relationship with plants. Left is an ant plant from South-East Asia, *Myrmecodia* sp., which provides an almost ready nest for ants which move in and protect the plant from its predators in return (cut away nest above). Right is a bull acacia thorn in Central America, which provides ants, *Pseudomyrmex ferruginea*, with special food nodules on its leaves, and the hollow spikes are very good as nests. In return the ants defend the plant against its enemies.

There are stories about fabled 'exploding ants' that would be easy to ignore except that they are true. Right: In Borneo, a small *Camponotus saundersi* ant has latched onto the face of its much larger attacker, a *Polyrhachis armata*, and mortally exploded the contents of its special sacs filled with noxious glue that will slowly kill the enemy. Left: Tiny *Crematogaster* ants, 0.3cm (0.1in), which can also explode the yellow sacs on their backs to spray attackers with toxic glue. They do not survive, but it immobilises the invader.

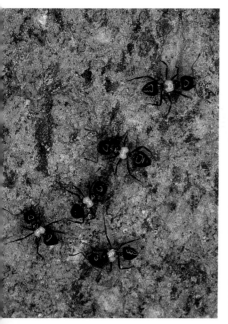

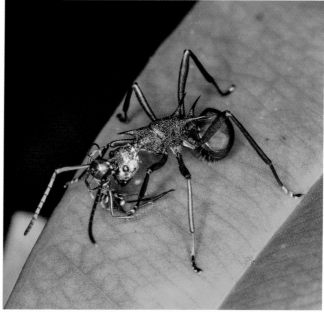

The fabled **army ants**, *Eciton* sp., of the tropical Americas are not quite like in the movies. Despite the amazing jaws on the soldier, most are hunters of other insects and do not specifically attack humans. The entire colony spends weeks resting in a pseudo nest made of their bodies, called a bivouac, or weeks foraging in columns of up to two million ants. All insect life in their path is eaten, except for the escaping insects which various ant-birds eat as they move with the ants.

To be truly afraid, it is not the army ants one needs to meet, but the **'bala' ant**, *Paraponera clavata*, of tropical Americas. Its other name is **bullet ant**, as the sting is considered the most painful of all insects, described as a throbbing, burning pain for 24 hours, acting so fast it evokes the feeling of being shot. 2.2cm (0.9in) long. One indigenous tribe in Brazil uses them in initiation rites!

Polyrhachis is one of the largest genera of **ants** with more than 600 mainly arboreal species. All have characteristic spines on top of the thorax. Their nests are often small and glued with silk. Top left the **golden tree ant**, and right, *Polyrhachis senilis*, both from Australia. Below left is a species from Indonesia, and below right is a clever mimic, a harmless Alydidae bug nymph. All around 0.6cm (0.25in) long.

Cephalotes basalis is one of the **turtle ants** or **gliding ants** in South America. They can glide between trees, where they nest in holes.

A soldier **seed ant**, *Pheidole* sp., whose huge jaws are used to crack some seeds in the underground nest. Plants can be helped by these ants, and some include a special ant food seed coating. 0.8cm (0.3in).

Africa has a similar legendary ant to the South American army ants. The **driver ants** or **siafu**, *Dorylus* sp., move in huge columns up to 50 million strong and attack everything in their path, including humans. Their bite is painful but not venomous, and because they don't let go they have been used as sutures to close wounds in places away from medical help. If they pass through a village or farm, they perform a pest eradication service, actually helping crops, although people have to move away as they pass. The soldier above is 1.8cm (0.7in) long.

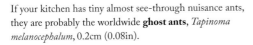

These are the leg bones of a large Cane Toad in Australia, stripped bare in one night by the well-named **meat ants**, *Iridomyrmex* sp., 1cm (0.4in)

Camponotus detritus is the only **ant** species to nest in the shifting dunes in the extreme sandy Namib Desert. 1cm (0.4in).

If your kitchen has tiny almost see-through nuisance ants, they are probably the worldwide **ghost ants**, *Tapinoma melanocephalum*, 0.2cm (0.08in).

It is rare to catch insects doing something as 'normal' as stopping for a drink. This **ant** in Borneo will take a big fill, and share it with ants back in the nest.

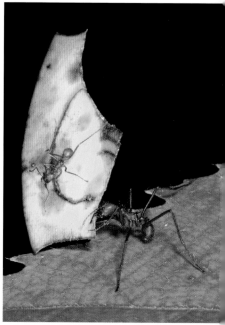

The **leafcutter ants**, genus *Atta*, are a constant source of wonder in tropical American forests. Their trails are hundreds of metres long, and their industry ceaseless. They have the most complex nests of all, with millions of ants in huge underground palaces, where special chambers grow a fungus on the leaves they bring. The fungus is the food of these agricultural ants. Above is a typical trail, and right a soldier carrying a leaf with a minor worker attached to scare off parasitic flies that try to parasitise the soldier. Below left a soldier uses its large mandibles to cut up a leaf. Soldier about 1.5cm (0.6in) long.

Below right: The last ant in this section is the largest. It is the soldier of the **carpenter ant**, *Camponotus gigas* from South-East Asia. Up to 2.8cm (1.2in) long, they are not aggressive and eat mainly honeydew.

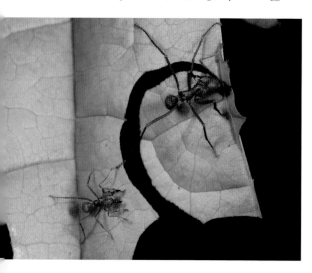

Bees are in the superfamily **Apoidea**. There are about 20,000 named species in seven families, with one shared character — of their hairs being branched. Most have no wasp-like waist, although they are essentially wasps. The honey bees, genus *Apis*, the small stingless bees like *Trigona* sp., and bumblebees have social organisation with complex colonies. But the majority of bees are solitary, making individual nests and burrows for their larvae, stocked with nectar or other plant products only. Bees are the most modern insect lineage, having co-evolved alongside flowering plants. Some families only have fossils going back 30 million years. The action of the whole group as pollinators is ecologically essential. *Apis* bees have stings, which unlike wasp and ant stings can only be used once, and kill the bee. Other bee species either tend not to sting, or cannot sting.

The **Apidae** contains the **honey bees**, **stingless bees**, **carpenter bees**, **cuckoo bees** and more, with about 6,000 named species.

Everyone knows the domestic Honey Bee, *Apis mellifera*, but here are two of its cousins. *Apis cerana* (left) is the **Asian Honey Bee**, a little bit smaller and with similar habits. On the right is the **Giant Honey Bee**, *Apis dorsata* from Asia, whose nests are often on cliff sides or high in tree branches, with no walls covering the huge honey combs. The latter is double the size of *A.mellifera*, and very dangerous when provoked.

The **stingless bees** are a group of several genera of small colony-building bees, which make a very nice thin honey. Their nests tend to be in tree holes, logs and even house walls, with characteristic resin entry tubes. Left: *Tetragona dorsalis* from Costa Rica, 0.8cm (0.3in) each. Right: *Hypotrigona* sp. from Ghana guarding the nest entrance, 0.6cm (0.25in) each.

More **stingless bees**. Top: *Trigona* sp. in a water lily in Australia, 0.5cm (0.2in) long. Below left: *Trigona atripes* collecting purple pollen from cat's whisker flowers in Malaysia, 0.6cm (0.25in) long. Below right: One of the largest of the group, *Meliponula bocandei*, from Ghana, where it is sometimes domesticated, 1cm (0.4in) long.

The **Xylocopinae** subfamily of **Apidae** are the **carpenter bees**. This contains about 500 species of large robust bees, most of which dig tunnel nests in soft living or decaying wood. Above left: *Xylocopa aruana* digging a fresh nest in Australia, 1.6cm (0.6in) long. Above right: Male *Xylocopa latipes*, a giant from Borneo, 3.2cm (1.3in) long. Note the little orange shape on its hind leg, which is a parasitic beetle larva. Below: *Xylocopa caffra* from South Africa, 3cm (1.2in) long.

Left: *Xylocopa coronata* from Indonesia, 2.5cm (1in) long. Right: *Xylocopa flavorufa* from South Africa, 2cm (0.8in) long.

Xylocopa varipes from Ghana. Like most **carpenter bees** its flight around flowering trees is very loud. 2.8cm (1.1in) long.

Xylocopa nigrita starting to dig a new nest hole in soft wood in Kenya. 2.2cm (0.9in) long.

The term **cuckoo bee** is applied to most species of the **Nomadinae** subfamily, and several genera of the **Apinae**, including one bumble bee. They have the same habit as the birds, laying their eggs in other bee nests where their larvae eat the pollen and the host larva. Left is *Thyreus nitidulus* from Australia, which uses blue-banded bee nests, and right is *Nomada leucophthalma* from Europe.

Another *Nomada* **cuckoo bee**. This genus has more than 850 species, and many look very wasp-like and have no pollen-collecting hairs, as they rely on other bees' work. From Indonesia.

Bombus (Psithyrus) vestalis. One subgenus of the *Bombus* bumble bee genus has evolved a cuckoo habit, invading other bumble bee nests and forcing the colony to look after it and its larvae. Europe.

The 250 species of bumble bees are the other group of social bees, after honey bees and stingless bees. They live in small, usually underground colonies, with a queen and workers. Left, *Bombus lapidarius* queen from Europe, 2.2cm (0.9in) long. Right, is *Bombus morrisoni* from North America, 1.6cm (0.6in) long.

The very common European **Buff-tailed Bumble Bee**, *Bombus terrestris*, is a good pollinator of many crops and has been exported to several other countries for the purpose. Colonies have up to 400 bees. 2cm (0.8in) long.

Amegilla is a genus in the **Apinae** subfamily with rounded, stout bees which make a characteristic high-pitched buzz with their very fast wing-beats. They dart around and hover briefly, visiting flowers for a second at a time. Left: One of the **blue-banded bees**, *A.cingulata*, 'robbing' a flower. When the flowers are too deep, these bees cut into the base and steal pollen without pollinating. Right: *A.zonata*.

Above: Male bees and many wasps often roost together at night. These *Amegilla zonata* males lock their jaws on a thin branch and remain immobile all night. The females (top right) have their nest holes to hide in. Australia, 1.2cm (0.5in) long.

The genus *Anthophora* contains around 450 species of solitary, fluffy, robust bees which make nests in the ground or in soil banks. Left is *A.plumipes* male from Europe, and right is a species from South Africa in its clay bank nest. Both about 1.2cm (0.5in) long.

The aptly named **teddy bear bee** from Australia is a fluffy male of *Amegilla bombiformis*. Male bees are sometimes larger and more hairy than the females. 1.5cm (0.6in) long.

Amegilla rhodoscymna from Australia, 1cm (0.4in) long.

Tetraloniella is a genus related to the *Anthophora* bees (previous page). Although not social, some species nest in huge groups when a suitable soil type is discovered. Shown here is a compacted dirt road in South Africa, with a busy digging female (right). 1.2cm (0.5in) long.

This *Amegilla* species in New Guinea is hovering in front of a flower it will 'rob' of its nectar. The mouthparts will pierce a hole at the base and bypass the flower front. It is an adaptation to obtain nutrients from flowers too deep to probe from the front. 1.4cm (0.6in) long.

A part of the **Apidae** family are known as **long-horned bees** for their long antennae. *Eucera nigrescens* is a European example. 1.2cm (0.5in) long.

Anthophora plumipes is aptly named the **long-tongued bee**. It can reach the nectar at the ends of these deep flowers it is hovering around. From Europe, 1cm (0.4in) long.

Allodapula variegata is a small bee with very large eyes. Its vision allows for foraging in near-dark conditions. Nests are in hollow stems. From South Africa, 0.8cm (0.3in) long.

Orchid bees are the most outrageously beautiful of all bees. They are a group of about 200 species in five genera within the **Apidae** family. They are rarely seen as they live high in trees in the tropical forests in Central and South America. The males pollinate specific orchids, and are very unusual for their habit of collecting scents from these and other sources, which they store and mix in special pockets in their enlarged rear legs. The theory is that they use these unique perfumes when wooing the females.

Orchid bees of three species of *Euglossa*, investigating strong scents attractive to their perfume-making needs. Costa Rica.

Some **orchid bees**, such as this large *Exaerete frontalis*, are parasitic on other orchid bee nests. Costa Rica, 1.5cm (0.6in) long.

Unknown *Euglossa* sp. **orchid bee** in Costa Rica. 0.8cm (0.3in) long.

Euglossa ignita hovering, a well-developed skill in **orchid bees**. Costa Rica, 1.2cm (0.5in) long.

One of the largest **orchid bees**, *Eurema mariana*, 1.6cm (0.6in) long.

Above: *Euglossa ignita*, showing two key features of **orchid bees**. The red arrow points to the special scent storage chamber in the enlarged back leg. What looks like a sting at the back is actually its very long mouthparts, held along and under the body, and used to pollinate very deep orchid flowers (see in top image also).

Above: *Euglossa sapphirina*, the **blue orchid bee**. Below: *Euglossa purpurea*.
Both from Costa Rica, and about 1cm (0.4in) long.

Halictidae, with more than 2,000 species, is the second-largest bee family. They are generally smaller, often metallic, and males often have yellow on their faces. Some have primitive social behaviour, and others are cuckoo bees invading other bee nests. The majority though are solitary, ground-nesting bees, provisioning their larvae with nectar and pollen.

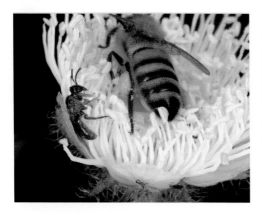

There is a belief that only the **Honey Bee**, *Apis mellifera* is a serious pollinator. Long before we domesticated this species, thousands of other bees, wasps and other insects were doing a grand job. The *Lasioglossum* sp. bee shown here is dwarfed by the Honey Bee, but quite up to the job of pollination.

Lasioglossum is the largest bee genus in the world, with more than 1,700 species. Here a Malaysian species is returning to its freshly dug communal nest with a huge pollen load. 0.8cm (0.3in).

This tiny *Homalictus* sp. from Australia is characteristically metallic. 0.6cm (0.2in) long.

This *Agapostemon* sp. is part of a group of metallic green bees, some of which nest in communal underground tunnels. From North America, 1cm (0.4in) long.

The very gold *Halictus (Seladonia) jucundus madecassus* from Madagascar. 1cm (0.4in) long.

Megalopta genalis, the **Night Bee**, has large eyes that are adapted to gather more than 20 times more light than those of normal bees, so it forages at night. It is also unusual as it can be either social or solitary. From Central America, 1cm (0.4in) long.

This **halictid bee** is 'robbing' flowers. It inserts its long sharp mouthparts into the bottom of the flower tube, stealing the nectar without saying returning the favour by pollinating the flower. Costa Rica, 1.5cm (0.6in) long.

The **Colletidae** has about 2,000 species, with the majority found in Australia and South America. Most bees collect pollen on special body hairs called the scopa, but many colletids do not have these, and instead carry pollen in an internal crop. Their brood cells are lined with a unique shiny smooth secretion, which has sometimes given the family the common name of **'polyester' bees** or **'plasterer' bees**. Most are solitary.

Colletes similis is a member of one of the two colletid genera found in Europe, with about 400 species. Most are solitary. 1cm (0.4in) long.

Hylaeus is a genus of about 500 species, known for carrying their pollen inside in a crop. *H.variegatus* is European. Many have species-unique patterns on the front of the head.

Hylaeus sanguinipictus is a specialist pollinator of the iconic Australian banksia flowers. 0.6cm (0.2in) long.

Male *(Paracolletes) Anthoglossa callander*, showing how often in the bee world, males are bigger and fluffier than the females. From Western Australia, 1.4cm (0.6in) long.

In the tropics, bees tend to live in open county or in the canopy of forests. One **colletid** genus, *Palaeorhiza*, are called **forest bees** as it is possible to find them in the gloomy shade of rainforests. At the top are two species from Australian jungles, below left is *Palaeorhiza exima* from Indonesia, and below right is a species from New Guinea. All about 1cm (0.4in).

Bees rarely mimic specific wasps for protection, but *Hyleoides* cf. *concinna* from Western Australia is one such mimic. 1cm (0.4in).

Hylaeus eligans from Western Australia, 1.2cm (0.5in) long.

The **Megachilidae**, with more than 4,000 species, are the **mason bees**, **resin bees**, **leafcutter bees**, **carder bees** and more. These names refer to their more complex nest building activities, above ground, and created from many substances, from resin to hair to leaves. Many species can be attracted to nesting boxes. Their bodies are often flattened rather than cylindrical, and the pollen-collecting hairs, the scopa, are on their abdomen, not on the legs as in honey bees.

Megachile frontalis from New Guinea, rubbing the pollen-collecting hairs on its abdomen against a pollen-rich flower. 1.5cm (0.6in) long.

A female *Coelioxys* sp., a **megachilid cuckoo bee**, in its odd upside-down locked-jaw sleeping stance. As it steals other bees' pollen stores, it has no collecting hairs (scopa) on the abdomen. New Guinea, 1.2cm (0.5in) long.

Members of the genus *Megachile*, with more than 1,500 species, are mainly **leafcutter bees**, with some **resin bees**. On the left: *M.bericetorum* is a European species, and on the right are resin nests constructed by a species of this genus in Western Australia.

Most **resin bees** obtain resins from plants, but this female **Fire-tailed Resin Bee**, *Megachile mystaceana*, in Australia is outside a stingless bee nest, stealing the resins they collected for their nest entrance. 1.2cm (0.5in) long.

A **leafcutter bee** in New Guinea carrying off a piece of leaf to line her brood cell with. The rounded holes are very characteristic for these bees, but they cut so fast it is rare to catch one in the act.

Camptopoeum friesei is a **resin bee** known as the **Salt Bee**, as it lives on salt-pans or the dried beds of soda lakes, making underground nests under the salt crust. From Austria. 1cm (0.4in) long.

Pachyanthidium cordatum is a **resin bee** from Kenya. It makes basket-like brood chambers out of plant hairs and resins. 1cm (0.4in) long.

This **leafcutter bee**, *Megachile* sp. (subgenus *Eutricharaea*), is from the dry Western Australian landscape. They create their nests in the ground, or in existing structures such as hollow branches or buildings. 1.2cm (0.5in) long.

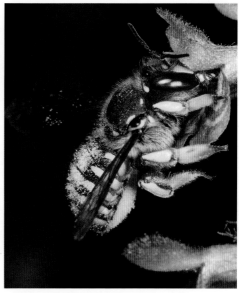

Coelioxys dispersa is a **leafcutter bee** in New Guinea. Although **megachilid** bees in general are not aggressive, the tail here is a reminder that they do possess stings. 1.2cm (0.5in) long.

Anthidium manicatum belongs to a genus known for their complex pot-like nests made from a mix of plant hair, resins from conifers and mud. The name **potter bee** often applies. From Europe. 1.2cm (0.5in) long.

Osmia aurulenta is a **mason bee** from Europe, here feeding on forget-me-not flowers (*Myosotis* sp.). Its nests are made from a complex mix of mud, clay, resin and plant hair. 1cm (0.4in) long.

A desert **leafcutter bee**, *Lithurgus* sp., inside a native hybiscus flower where it will pollinate, feed and spend the night. 1.2cm (0.5in) long.

Members of the family **Andrenidae** are known as **mason bees**. About 3,000 species make their nests in the ground, in complex tunnels with many branches, and with a larva at the end of each branch. They prefer open ground and therefore many are found in arid areas of the world.

Andrena aegyptiaca is one of more than 1,300 species of this genus. It makes complex tunnels in hard sandy soils, with multiple brood chambers, and is found in Europe and North Africa. 1cm (0.4in) long.

Andrena fulva, the **Tawny Mining Bee**, is common throughout Europe. Many females may make branching ground tunnels close to each other, but they are not communal. 1cm (0.4in) long.

Andrena cineraria is the **Ashy Mining Bee**, found across Europe and north Asia. It makes branched tunnels in grassy areas.

Species of the red **oil bee** genus *Haplomelitta* are restricted to Africa, which is a key location for **Melittidae** evolution in general. 1.2cm (0.5in) long.

The **Melittidae**, the **oil-collecting bee** family, has only about 200 species. They are known for having a close co-evolutionary relationship with the plants they pollinate, sometimes visiting only one kind of flower. This is because some feed the larvae on pollen and nectar, as well as oils from a particular flower source. This is made into a ball buried in waterproofed, underground brood cells.

Macropis fulvipes, an **oil bee** restricted to flowers of the genus *Lasimachia*, loosestrife and pimpernel. These provide the extra oils this bee uses in the larval feed. Europe, 1.2cm (0.5in).

Melitta haemorrhoidalis resting in a *Campanula* flower. On dull and cold days, many bees will sometimes rest in a sheltered spot. Found in Europe, 1cm (0.4in) long.

GLOSSARY

Abdomen
the rear of the three main body parts of an insect, containing the gut and reproductive parts.

Appendage
any segmented body part attached to the main body, such as the legs and antennae.

Apterous
wingless.

Arthropod
any member of the phylum Arthropoda, which all share a body with an external skin-like skeleton, with jointed body and limb segments.

Brachypterous
with shorter than normal wings.

Caste
in social or colonial insects, any distinct group performing a distinct job, such as the workers and soldiers of ants and termites.

Cerci
a pair of jointed appendages at the tail-end of some insects, such as cockroaches, which has extra-sensory functions.

Chitin
the very resilient, plastic-like, polysachharide substance, forming the exoskeleton of all arthropods.

Chrysalis
the pupal stage of butterflies and moths, usually covered in silk among the moths, and exposed and often bright in butterflies.

Clubbed
describing antennae which end in a bulbous larger segment.

Colony
a group of nest-sharing social insects.

Complete metamorphosis
the growth cycle where the young (the larvae) have a different form and lifestyle to the adult, with a pupa stage between the two. Like the caterpillar-pupa-adult stages of butterflies.

Cosmopolitan
very widespread, found basically everywhere.

Coxa (Coxae)
the first segment of insect legs, sometimes plate-like and fixed to the main body.

Crypsis (Cryptic)
resembling and blending with the surface it sits on. Camouflaged insects are cryptic.

Cuticle
the hard, jointed outer skeleton, or 'skin', of insects.

Detritus
broken up and usually decaying organic matter.

Dorsal
the upper or top surface.

Ectoparasite
a parasite living on the outside of the host's body, such as a tick.

Elytra
the hard forewings of beetles, which serve as a cover to protect the membranous flying hind-wings.

Endemic
restricted in distribution to a particular region.

Endoparasite
a parasite living inside the body of its host or 'victim'.

Eusocial
the most organised form of insect societies, with separate castes, including reproductive and sterile forms, and a defended nest with up to several generations of workers. Mainly ants, termites and bees.

Exoskeleton
the external skeleton of arthropods such as insects.

Femur
the third, most powerful, segment of the five-segmented legs of insects.

Filiform
thread-like, describing antennae which are made of elongate thread-like parts.

Frass
plant fragments made by wood-boring insects, usually mixed with excreta.

Frenulum
a set of curled bristles on the hind-wings of many moths which link the wings in flight, acting much like velcro.

Gall
a tumorous growth on plants, often bulbous, caused by specific irritant chemicals injected by certain insects whose larvae develop inside the growth, feeding on special softer tissue.

Gaster
the main part of the abdomen on wasps, behind the characteristic thinner 'waist'.

Glossa
the mouthparts of bees, their 'tongue', used in extracting pollen and other substances.

Halteres
the club-like structures all flies have in place of the second pair of wings, used as balancing tools for finer flight control.

Hemelytra
the partly hardened and partly membranous protective forewings of true bugs (Hemiptera) which can still be used in flight.

Honeydew
the sweet secretions of aphids, planthoppers and some other insects, which attract ants, who in turn often protect these 'farmed' insects.

Hygroscopic
a substance or structure which repels water.

Hyperparasite
a parasite which parasitises another parasite.

Imago
the adult, sexually mature stage of insects.

Incomplete metamorphosis
the growth cycle where the young – nymphs – are similar in appearance to, and share the lifestyle of, the adults. They develop gradually through successive stages, and become an adult without going through a pupa stage.

Inquiline
an animal, mainly insects, living in the nest of other species, mainly of ants and termites. A 'social' parasite.

Instar
one of the stages of immature insects, either of larvae or nymphs.

Kleptoparasite
a female wasp or bee which seeks out other species' nests and raids their food supply to raise her own offspring, From the Greek word for thief, 'kleptes'.

Labial palps

one of two sets of jointed palps on the side of the mouth of insects, used to manipulate food.

Labrum

A plate or flap at the front of the head, much like an 'upper lip', as it often partly covers the mouthparts.

Lamellate

fan-like, describing antennae with the last few segments like an opening fan.

Larva

the immature stage of insects which undergo full metamorphosis from egg, to larvae, to pupa, to adult. Larvae grow by shedding or moulting their exoskeletons usually about four to six times.

Larviform

in some insect orders, the females may still look like the wingless larva after metamorphosis, but is actually an adult ready to breed.

Lateral

pertaining to the sides.

Mandibles

the upper, chewing pair of hard mouthparts, sometimes modified into other structures.

Mandibulate

with biting or chewing mouthparts.

Maxillae

the second, partly internal, pair of mouthparts in insects.

Maxillary palps

one of two sets of jointed palps, on the side of the mouths of insects, used to manipulate food, and behind the labial palps.

Membranous

pertaining to wings, the usually transparent flying wings, such as in dragonflies.

Moniliform

bead-like, describing antennae composed of spherical segments.

Moult (molt)

to shed the outer skeleton or 'skin' in the process of growth.

Nymph

the immature stage of insects which do not undergo full metamorphosis, but grow through stages that look like the adult, minus wings and reproductive parts, such as true bugs.

Ocellus (ocelli)

a simple single-lensed eye often present in a pattern of three on top of the head of many insect groups.

Ovipositor

a tubular or sword-like egg-laying apparatus, concealed in some insects, or large and robust in crickets, or extra long and thin in some wasps.

Parthenogenesis

reproduction from unfertilised eggs, practiced by a few insects, such as some stick insect and aphids.

Pedicel

the two segmented 'waist' of an ant.

Petiole

the narrow, usually one segmented 'waist' of wasps.

Phytophagous

feeding on plants.

Plastron

a layer of fine hair on the underside of some aquatic insects, which traps a layer of air, often looking like quicksilver.

Proboscis

extended mouthparts which are often modified into a tube for sucking, like the straight proboscis of mosquitoes, or the coiled one of moths and butterflies.

Prolegs
tiny sucker-like, flat, false legs, on the abdominal segments behind the real legs, in caterpillars of moths and butterflies.

Pronotum
the upper, dorsal, surface of the first segment of the thorax. For example, in beetles it is the segment seen from above between the head and winged body.

Pterygote
winged.

Pubescent
covered in fine hair, downy.

Pupa
the stage between the larva and the adult in insects which undergo full metamorphosis, inside it is rearranging tissues into the new form.

Raptorial legs
grasping, often spined forelegs used to capture and hold prey in insects like mantids, and evolved in several other small groups.

Receptor
any organ for receiving input from the environment around the animal. In insects it can be the antennae, body hairs, palps, and more.

Resilin
a rubber-like very elastic protein found in some leg and other joins of insects. When stretched it stores energy and if released rapidly, provides a huge boost to actions like jumping, as in fleas.

Rostrum
the hard, beak-like, tubular mouthparts of true bugs (Hemiptera) used for sucking plant sap or the contents of prey.

Saprophagous
feeding on decaying organic matter.

Scale
modified hair, attached to hair sockets, which is broad and flat, like the scales of butterflies.

Sclerite
any of the body plates making up the segments of the exoskeleton of insects.

Segment
one part of the many segmented bodies of insects, which can be a segment of the head, thorax, abdomen, legs or other appendages.

Semi-social
insect colonies where females of one generation cooperate to care for the young, and partly divide labour into reproductive and worker. A step in the evolution of full social behaviour.

Serrate
saw-like or tooth-like, describing antennae with such shaped segments.

Seta (setae)
stiff, bristle-like body hairs.

Social
one of several systems adapted by group-living insects, from semi-social to complex multi-generational eusocial societies.

Soldier
a member of the worker caste in ants and termites, adapted for defence with extra strength, size and usually larger mandibles.

Spiracle
one of many external openings, literally tiny holes along the body segments, into the breathing system of insects.

Sting
the modified ovipositor of some wasps, bees and ants, which does not lay eggs but injects venom.

Stridulation

production of sound by rubbing two parts of the insect body, usually a toothed file on legs and/or wings. Grasshopper and cricket sound mechanism.

Symbiosis

a relationship between two organisms where both get benefit.

Tarsus (tarsi)

the last segments of the insect leg, the 'foot', usually three- to five-segmented, thin and ending on a pair of claws.

Tegmen (tegminal)

the toughened, leathery forewings of several orders of insects, especially mantids, cockroaches and grasshoppers and crickets.

Thorax

the middle of the three main body divisions of insects, between the head and abdomen. The three pairs of legs and one or two pairs of wings are attached here.

Tibia

the fourth segment of an insect leg, between the powerful femur, and the tarsi, or 'feet'.

Trochanter

the second segment of an insect leg, before the femur.

Tymbal

the sound-producing, drum-like membrane of a cicada.

Vector

an insect which transmits disease between hosts, like some mosquito species or flies.

Venation

the pattern of veins in an insect wing, used in describing species.

Venom

a poison used by spiders, and among insects, the wasps, injected into prey or attacking predators.

Ventral

the lower or underside surface of body parts.

Vestigial

only partly developed, small and non-functional, like shortened non flying wings in some insects.

Viviparous

bearing live young, no external egg stage. Aphids and some flies are main examples.

Warning patterns

very garish, contrasting patterns found in many insects which store poisons and are dangerous for predators to try and eat.

FURTHER READING – BOOKS AND WEB LINKS

BOOKS

Brock, Paul D. 2017. *A Photographic Guide to Insects of Southern Europe and the Mediterranean*, Pisces Publications.

Chinery, Michael. 2009. *British Insects: A photographic guide to every common species*, Collins.

Gooderham, J and Tsyrlin, E. 2012. *The Waterbug Book*, CSIRO Publishing.

Hoskins, Adrian. 2015. *Butterflies of the World*. Reed New Holland.

Kaufman, K and Eaton, Eric R 2007. *Kaufman Field Guide to Insects of North America*, Kaufman Field Guides, Turtleback.

Marshall, Stephen. A. 2018. *Beetles: The Natural History and Diversity of Coleoptera*, Firefly Books.

Marshall, Stephen. A. 2017. *Insects: Their Natural History and Diversity*, Firefly Books.

Marshall, Stephen. A. 2012. *Flies: The Natural History and Diversity of Flies*. Firefly Books.

Naskrecki, P. 2005. *The Smaller Majority*, Belknap Press of Howard University Press.

Naskrecki, P. and Wilson, E.O. 2017. *Hidden Kingdom: The Insect Life of Costa Rica*, Zona Tropica Publications.

The Natural History Museum, London, 1995. *Megabugs, The Natural History Museum Book of Insects*, Carlton Books.

Zborowski, Paul 2016. Bloodsuckers, Young Reed.

Zborowski, Paul 2010. *Can You Find Me?, Nature's Hidden Creatures*, Young Reed.

Zborowski, Paul, and Storey, Ross. 2017. *A Field Guide to Insects in Australia*, 4th Edition, Reed New Holland.

THE WEB

Most museums around the world have great insect sites - look for them locally, and visit the museums if you can.

Australian Museum insect pages: australianmuseum.net.au/insects

University of Florida Book of Insect Records: entomology.ifas.ufl.edu/walker/ufbir

North American insect online guides:
bugguide.net/node/view/15740
www.insectidentification.org

Natural History Museum in London, insect site: www.nhm.ac.uk/our-science/departments-and-staff/life-sciences/insects.html

general insect resources sites:
www.insects.org
www.insectimages.org/index.cf
www.si.edu/spotlight/buginfo/incredbugs

INDEX

Names in Bold are insect families.

IMAGE CREDITS

ABOUT THE AUTHOR

Paul Zborowski is an entomologist and photographer who spends his time researching and illustrating invertebrates around the world. He has a passion for presenting the macro details of nature to a wide audience and has written and illustrated 15 nature books using the resources of his image bank, www.close-up-photolibrary.com. He is a field-based scientist, spending long periods undertaking research in locations all around the world, and is also sometimes a teacher, both of science and photography.

First published in 2019 by Reed New Holland Publishers
Sydney • Auckland

Level 1, 178 Fox Valley Road, Wahroonga, NSW 2076, Australia
5/39 Woodside Avenue, Northcote, Auckland 0627, New Zealand

newhollandpublishers.com

A record of this book is held at the National Library of Australia.

ISBN 978 1 92554 609 5

Group Managing Director: Fiona Schultz
Publisher and Project Editor: Simon Papps
Designer: Andrew Davies
Production Director: Arlene Gippert
Printer: Toppan Leefung Printing Limited

10 9 8 7 6 5 4 3 2 1

Keep up with Reed New Holland:
 ReedNewHolland
 @reednewholland

US $39.99